# *Distress Identification Manual for the Long-Term Pavement Performance Project*

**SHRP-P-338**

**STRATEGIC HIGHWAY RESEARCH PROGRAM**
National Research Council

Washington, DC   1993

Publication No. SHRP-P-338
ISBN 0-309-05271-8
Contract P-001
Product No. 5016


**Program Managers:**     Neil F. Hawks
                          T. Paul Teng

**Project Managers:**     William Yeadon Bellinger
                          Richard Ben Rogers

**Program Area Secretary:**  Cynthia Baker

**Copyeditor:**            Katharyn L. Bine Brosseau

**Graphic Designer:**      Linda C. Humphrey


May 1993

---

**KEY WORDS**

**asphalt concrete**
**distress**
**pavement performance**
**portland cement concrete**

---

Strategic Highway Research Program
2101 Constitution Avenue NW
Washington, DC  20418

(202) 334-3774

# Contents

## 1

# DISTRESSES FOR PAVEMENTS WITH ASPHALT CONCRETE SURFACES / 5

## 2

# DISTRESSES FOR PAVEMENTS WITH JOINTED PORTLAND CEMENT CONCRETE SURFACES / 37

**3**

# DISTRESSES FOR PAVEMENTS WITH CONTINUOUSLY REINFORCED CONCRETE SURFACES / 63

**A**

# MANUAL FOR DISTRESS SURVEYS / 91

**B**

# MANUAL FOR DIPSTICK PROFILE MEASUREMENTS / 127

**C**

# MANUAL FOR FAULTMETER MEASUREMENTS / 143

# List of Figures

vi

## List of Tables

In 1987, the Strategic Highway Research Program began the largest and most comprehensive pavement performance test in history—the Long-Term Pavement Performance (LTPP) program. During the program's 20-year life, highway agencies in the United States and 15 other countries will collect data on pavement condition, climate, and traffic volumes and loads from more than a thousand pavement test sections. That information will allow pavement engineers to design better, longer-lasting roads.

*The Distress Identification Manual for the Long-Term Pavement Performance Project* was developed to provide a consistent, uniform basis for collecting distress data for the LTPP program. It will allow states and others to provide accurate, uniform, and comparable information on the condition of LTPP test sections. The manual will also prove useful to individuals and agencies intending to interpret LTPP data or correlate LTPP findings with their own research efforts.

Although developed as a tool for the LTPP program, the manual has broader application. It provides a common language for describing cracks, potholes, rutting, spalling, and other pavement distresses being monitored by the LTPP program. As a "distress dictionary," the manual will improve inter- and intra-agency communication and lead to more uniform evaluations of pavement performance.

The manual is divided into three sections, each focusing on a particular type of pavement: (1) asphalt concrete surfaced, (2) jointed portland cement concrete, and (3) continuously reinforced portland cement concrete. Each distress is clearly labeled, described, and illustrated.

**Foreword**

## GUIDANCE TO USERS

### Researchers

If you are monitoring LTPP or other test sections, please follow the guidelines in Appendix A ("Manual for Distress Surveys") to ensure the data collected will be comparable to other LTPP data. Sample data collection sheets are included in the appendix. As you evaluate a section of roadway, keep the manual handy; to determine the type and severity of distress, find the definition and illustration that best matches the pavement section being surveyed.

Appendices B and C describe how to use the Dipstick profile measurement device and the Georgia Digital Faultmeter.

For more assistance in the identification of pavement distress, contact the Federal Highway Administration's Long-Term Pavement Performance Division or the National Highway Institute about their distress identification workshop.

### Other Users

As a pavement distress dictionary, the manual will improve communications within the pavement community by fostering more uniform and consistent definitions of pavement distress. Highway agencies, airports, parking facilities, and others with significant investment in pavements will benefit from the adoption of a standard distress language.

Colleges and universities will use the manual in highway engineering courses. It serves also as a valuable training tool for highway agencies. Now, for example, when a distress is labeled "high severity fatigue cracking," it is clear exactly what is meant. Repairs can be more efficiently planned and executed, saving the highway agency crew time and money.

Although not specifically designed as a pavement management tool, the *Distress Identification Manual* can play an important role in a state's pavement management program by ridding reports of inconsistencies and variations caused by a lack of standardized terminology. Most pavement management programs do not need to collect data at the level of detail and precision required for the LTPP program, nor are the severity levels used in the manual necessarily appropriate for all pavement management situations. Thus you may choose to modify the procedures (but not the definitions) contained in the manual to meet your specific needs, taking into account the desired level of detail, accuracy and timeliness of information, available resources, and predominant types of distress within the study area.

## FOR MORE INFORMATION

In 1992, the U.S. Department of Transportation's Federal Highway Administration assumed responsibility for coordinating the LTPP activities and collecting and analyzing the LTPP data. The National Research Council's Transportation Research Board maintains the LTPP database.

For information on the LTPP program, contact the Federal Highway Administration, Office of Research and Development, LTPP Division, Turner-Fairbank Highway Research Center, 6300 Georgetown Pike, McLean, VA 22101 (telephone: 703/285-2355).

For information on the LTPP database, contact Michael Morales, Transportation Research Board, 2101 Constitution Ave., N.W., Washington DC 20418 (telephone: 202/334-2239).

**Preface**

The research described herein was supported by the Strategic Highway Research Program (SHRP). SHRP is a unit of the National Research Council that was authorized by section 128 of the Surface Transportation and Uniform Relocation Assistance Act of 1987.

SHRP was created as a five-year program. The goals of SHRP's Long-Term Pavement Performance (LTPP) program, however, required an additional fifteen years of research. To meet these goals, LTPP was transferred from SHRP to the Federal Highway Administration (FHWA) of the U.S. Department of Transportation on July 1, 1992, in accordance with the mandate of the Intermodal Surface Transportation Efficiency Act of 1991.

The first SHRP *Distress Identification Manual for the Long-Term Pavement Performance Studies* (1987) was authored by Kurt D. Smith, Michael I. Darter, and Kathleen T. Hall of ERES Consultants, Inc., Champaign, Illinois, and J. Brent Rauhut of Brent Rauhut Engineering, Austin, Texas. Support for that work was provided by the FHWA under Contract No. DTFH61-85-C-0095 as part of a "transition plan" to support planned implementation of LTPP monitoring, pending SHRP funding authorization by Congress.

A second version, the *Distress Identification Manual for the Long-Term Pavement Performance Studies* (1990), was developed by Karen Benson of the Texas Research and Development Foundation (TRDF), Austin, Texas, and Humberto Castedo and Dimitrios G. Goulias of the University of Texas at Austin, Center for Transportation Research (CTR), with guidance and support from W. R. Hudson of the University of Texas. Support for the revision work was provided by SHRP as a part of Contract SHRP-87-P001.

This third version was developed by John S. Miller, P.E., of Law Engineering, Inc., Richard Ben Rogers, P.E., Texas State Department of Transportation, and Gonzalo R. Rada, Ph.D., P.E., of Law Engineering, with guidance and support from William Yeadon Bellinger, P.E., of the FHWA. Guidance and advice was also provided by the Distress Identification Manual Expert Task Group (ETG). See cover 3 for a listing of ETG members. Support for the revision was provided by the FHWA as a part of Contract DTFH61-92-C-00134.

Doug Frith and Dan Bryant, SHRP Western Regional Office, Jerry Daleiden, SHRP Southern Regional Office, Brian Nieuwenhuis, SHRP North Central Regional Office, and Andrew Brigg, SHRP North Atlantic Regional Office ably reviewed this revision of the manual.

Valuable information, material, and technical support were provided by: the National Association of Australian State Road Authorities; Ontario Ministry of Transportation and Communications; American Public Works Association; the Asphalt Institute; the Kentucky Transportation Cabinet; the Michigan Department of Transportation; the Mississippi State Highway Department; the Missouri Highway and Transportation Department; the North Carolina Department of Transportation; the Pennsylvania Department of Transportation; the Texas Department of Transportation; and the Washington State Department of Transportation.

# Distress Identification Manual for the Long-Term Pavement Performance Project

Accurate, consistent, and repeatable distress evaluation surveys can be performed by using the *Distress Identification Manual for the Long-Term Pavement Performance Project.* Color photographs and drawings illustrate the distresses found in three basic pavement types: asphalt concrete-surfaced; jointed (plain and reinforced) portland cement concrete; and continuously reinforced concrete. Drawings of the distress types provide a reference to assess their severity. Methods for measuring the size of distresses and for assigning severity levels are given. The manual also describes how to conduct the distress survey, from obtaining traffic control to measuring the cracks in the pavement. Sample forms for recording and reporting the data are included. The manual also tells how to calibrate and operate profile and fault measurement devices.

**Abstract**

This section covers asphalt concrete (AC) surfaced pavements, including AC overlays on either asphalt concrete or portland cement concrete pavements. Each of the distresses has been grouped into one of the following categories:

- **A.** Cracking
- **B.** Patching and Potholes
- **C.** Surface Deformation
- **D.** Surface Defects
- **E.** Miscellaneous Distresses

Table 1 summarizes the various types of distress and unit of measurement. Some distresses also have defined severity levels.

**1**

**DISTRESSES FOR PAVEMENTS WITH ASPHALT CONCRETE SURFACES**

| **TABLE 1.** Asphalt Concrete-Surfaced Pavement Distress Types | | |
|---|---|---|
| **DISTRESS TYPE** | **UNIT OF MEASURE** | **DEFINED SEVERITY LEVELS?** |
| **A.** Cracking / page **7** | | |
| 1. Fatigue Cracking | Square Meters | Yes |
| 2. Block Cracking | Square Meters | Yes |
| 3. Edge Cracking | Meters | Yes |
| 4a. Wheel Path Longitudinal Cracking | Meters | Yes |
| 4b. Non-Wheel Path Longitudinal Cracking | Meters | Yes |
| 5. Reflection Cracking at Joints | | |
|     Transverse Reflection Cracking | Number, Meters | Yes |
|     Longitudinal Reflection Cracking | Meters | Yes |
| 6. Transverse Cracking | Number, Meters | Yes |
| **B.** Patching and Potholes / page **19** | | |
| 7. Patch/Patch Deterioration | Number, Square Meters | Yes |
| 8. Potholes | Number, Square Meters | Yes |
| **C.** Surface Deformation / page **25** | | |
| 9. Rutting | Millimeters | No |
| 10. Shoving | Number, Square Meters | No |
| **D.** Surface Defects / page **29** | | |
| 11. Bleeding | Square Meters | Yes |
| 12. Polished Aggregate | Square Meters | No |
| 13. Raveling | Square Meters | Yes |
| **E.** Miscellaneous Distresses / page **33** | | |
| 14. Lane-to-Shoulder Dropoff | Millimeters | No |
| 15. Water Bleeding and Pumping | Number, Meters | No |

This section includes the following distresses:

1. Fatigue Cracking
2. Block Cracking
3. Edge Cracking
4a. Longitudinal Cracking - Wheel Path
4b. Longitudinal Cracking - Non–Wheel Path
5. Reflection Cracking at Joints
6. Transverse Cracking

Measurement of crack width is illustrated in Figure 1. Figure 2 depicts the effect on severity level of a crack, in this case block cracking, due to associated random cracking.

**Cracking**

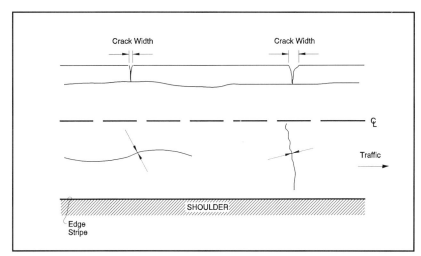

**FIGURE 1**
**Measuring Crack Width in Asphalt Concrete-Surfaced Pavements**

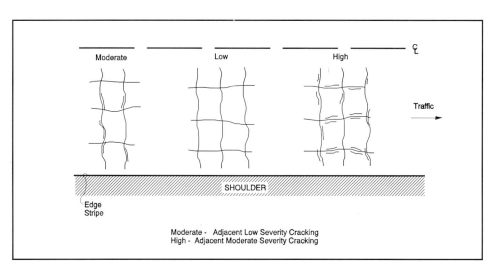

**FIGURE 2**
**Effect on Severity Level of Block Cracking due to Associated Random Cracking**

**FATIGUE CRACKING**

**Description**

Occurs in areas subjected to repeated traffic loadings (wheel paths).

Can be a series of interconnected cracks in early stages of development. Develops into many-sided, sharp-angled pieces, usually less than 0.3 m (1 ft) on the longest side, characteristically with a chicken wire/alligator pattern, in later stages.

Must have a quantifiable area.

**Severity Levels**

**LOW**
An area of cracks with no or only a few connecting cracks; cracks are not spalled or sealed; pumping is not evident.

**MODERATE**
An area of interconnected cracks forming a complete pattern; cracks may be slightly spalled; cracks may be sealed; pumping is not evident.

**HIGH**
An area of moderately or severely spalled interconnected cracks forming a complete pattern; pieces may move when subjected to traffic; cracks may be sealed; pumping may be evident.

**How to Measure**

Record square meters (square feet) of affected area at each severity level.

If different severity levels existing within an area cannot be distinguished, rate the entire area at the highest severity present.

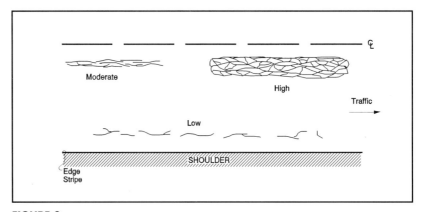

**FIGURE 3**
ACP 1. Fatigue Cracking

**ASPHALT CONCRETE SURFACES**

8

**FIGURE 4**
ACP 1. Chicken Wire/Alligator Pattern
Cracking Typical in Fatigue Cracking

**FIGURE 5**
ACP 1. Moderate Severity Fatigue Cracking

**FIGURE 6**
ACP 1. High Severity Fatigue Cracking

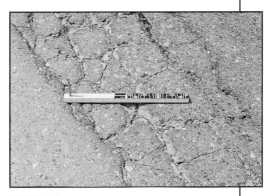

**FIGURE 7**
ACP 1. High Severity Fatigue Cracking
With Spalled Interconnected Cracks

Cracking

# BLOCK CRACKING

## Description

A pattern of cracks that divides the pavement into approximately rectangular pieces. Rectangular blocks range in size from approximately 0.1 sq. m to 10 sq. m (1 sq. ft to 100 sq. ft).

## Severity Levels

### LOW

Cracks with a mean width ≤ 6 mm (0.25 in.); or sealed cracks with sealant material in good condition and with a width that cannot be determined.

### MODERATE

Cracks with a mean width > 6 mm (0.25 in.) and ≤ 19 mm (0.75 in.); or any crack with a mean width ≤ 19 mm (0.75 in.) and adjacent low severity random cracking.

### HIGH

Cracks with a mean width > 19 mm (0.75 in.); or any crack with a mean width ≤ 19 mm (0.75 in.) and adjacent moderate to high severity random cracking.

## How to Measure

Record square meters (square feet) of affected area at each severity level.

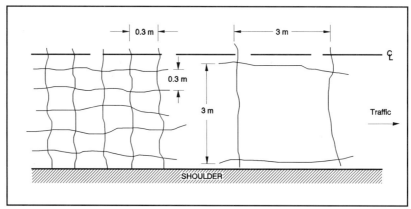

**FIGURE 8**
**ACP 2. Block Cracking**

**FIGURE 9**
**ACP 2. Moderate Severity Block Cracking**

**FIGURE 10**
**ACP 2. High Severity Block Cracking**

**ASPHALT CONCRETE SURFACES**

# EDGE CRACKING

## Description

Applies only to pavements with unpaved shoulders.

Crescent-shaped cracks or fairly continuous cracks which intersect the pavement edge and are located within 0.6 m (2 ft) of the pavement edge, adjacent to the shoulder. Includes longitudinal cracks outside of the wheel path and within 0.6 m (2 ft) of the pavement edge.

## Severity Levels

### LOW
Cracks with no breakup or loss of material.

### MODERATE
Cracks with some breakup and loss of material for up to 10% of the length of the affected portion of the pavement.

### HIGH
Cracks with considerable breakup and loss of material for more than 10% of the length of the affected portion of the pavement.

## How to Measure

Record length in meters (feet) of pavement edge affected at each severity level. The combined quantity of edge cracking cannot exceed the length of the section.

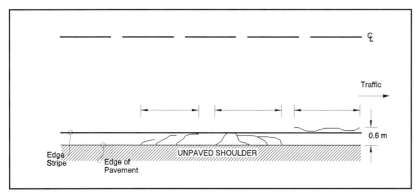

**FIGURE 11**
**ACP 3. Edge Cracking**

**FIGURE 12**
**ACP 3. Low Severity Edge Cracking**

Cracking

11

# 4. LONGITUDINAL CRACKING

## Description

Cracks predominantly parallel to pavement centerline.  Location within the lane (wheel path versus non-wheel path) is significant.

## Severity Levels

### LOW
A crack with a mean width ≤ 6 mm (0.25 in.); or a sealed crack with sealant material in good condition and with a width that cannot be determined.

### MODERATE
Any crack with a mean width > 6 mm (0.25 in.) and ≤ 19 mm (0.75 in.); or any crack with a mean width ≤ 19 mm (0.75 in.) and adjacent low severity random cracking.

### HIGH
Any crack with a mean width > 19 mm (0.75 in.); or any crack with a mean width ≤ 19 mm (0.75 in.) and adjacent moderate to high severity random cracking.

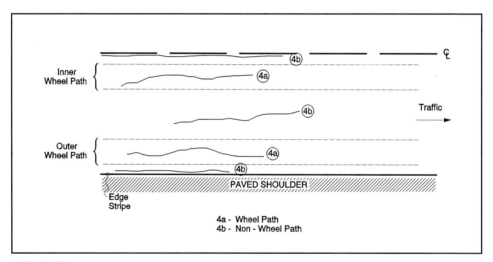

**FIGURE 13**
ACP 4. Longitudinal Cracking

## How to Measure

Record separately:

**4A. WHEEL PATH LONGITUDINAL CRACKING**

Record the length in meters (feet) of longitudinal cracking within the defined wheel paths at each severity level.

Record the length in meters (feet) of longitudinal cracking with sealant in good condition at each severity level.

**4B. NON-WHEEL PATH LONGITUDINAL CRACKING**

Record the length in meters (feet) of longitudinal cracking not located in the defined wheel paths at each severity level.

Record the length in meters (feet) of longitudinal cracking with sealant in good condition at each severity level.

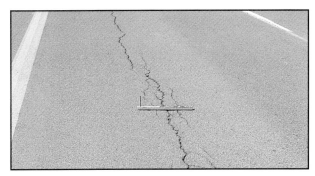

**FIGURE 14**
ACP 4a. Moderate Severity Longitudinal Cracking in the Wheel Path

**FIGURE 15**
ACP 4b. High Severity Longitudinal Cracking Not in the Wheel Path

Cracking

13

**5**

# REFLECTION CRACKING AT JOINTS

### Description

Cracks in asphalt concrete overlay surfaces that occur over joints in concrete pavements.

Note: Knowing the slab dimensions beneath the asphalt concrete surface helps to identify reflection cracks at joints.

### Severity Levels

**LOW**
An unsealed crack with a mean width ≤ 6 mm (0.25 in.); or a sealed crack with sealant material in good condition and with a width that cannot be determined.

**MODERATE**
Any crack with a mean width > 6 mm (0.25 in.) and ≤ 19 mm (0.75 in.); or any crack with a mean width ≤ 19 mm (0.75 in.) and adjacent low severity random cracking.

**HIGH**
Any crack with a mean width > 19 mm (0.75 in.); or any crack with a mean width ≤ 19 mm (0.75 in.) and adjacent moderate to high severity random cracking.

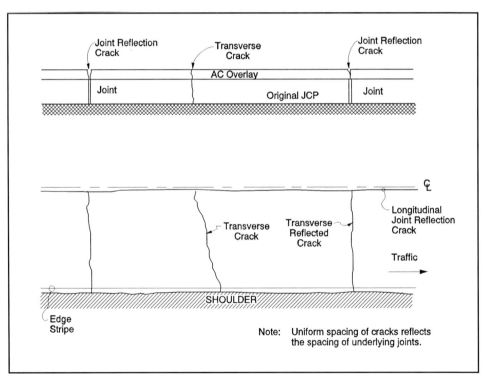

**FIGURE 16**
ACP 5. Reflection Cracking at Joints

14

## How to Measure

### TRANSVERSE REFLECTION CRACKING:

Record the number of transverse reflection cracks at each severity level. Rate each transverse reflection crack at the highest severity level present for at least 10% of the total length of the crack.

Record length in meters (feet) of transverse reflection cracks, assigned to the severity level of the crack.

Record length in meters (feet) of transverse cracks with sealant in good condition at each severity level.

Note: The length recorded is the total length of the well-sealed crack and is assigned to the severity level of the crack. Record only when the sealant is in good condition for at least 90% of the length of the crack.

### LONGITUDINAL REFLECTION CRACKING:

Record length in meters (feet) of longitudinal reflection cracking at each severity level.

Record the length in meters (feet) of longitudinal reflection cracking with sealant in good condition at each severity level.

**FIGURE 17**
**ACP 5. High Severity Reflection Cracking at Joints**

## 6 | TRANSVERSE CRACKING

### Description

Cracks that are predominantly perpendicular to pavement centerline, and are not located over portland cement concrete joints.

### Severity Levels

**LOW**

An unsealed crack with a mean width ≤ 6 mm (0.25 in.); or a sealed crack with sealant material in good condition and with a width that cannot be determined.

**MODERATE**

Any crack with a mean width > 6 mm (0.25 in.) and ≤ 19 mm (0.75 in.); or any crack with a mean width ≤ 19 mm (0.75 in.) and adjacent low severity random cracking.

**HIGH**

Any crack with a mean width > 19 mm (0.75 in.); or any crack with a mean width ≤ 19 mm (0.75 in.) and adjacent moderate to high severity random cracking.

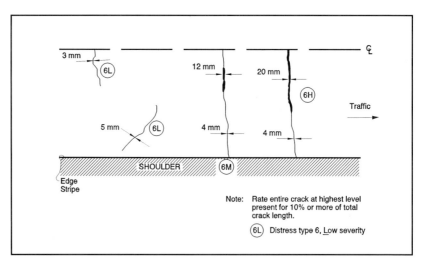

**FIGURE 18**
ACP 6. Transverse Cracking

## How to Measure

Record number and length of transverse cracks at each severity level. Rate the entire transverse crack at the highest severity level present for at least 10% of the total length of the crack. Length recorded, in meters (feet), is the total length of the crack and is assigned to the highest severity level present for at least 10% of the total length of the crack.

Also record length in meters (feet) of transverse cracks with sealant in good condition at each severity level.

Note: The length recorded is the total length of the well sealed crack and is assigned to the severity level of the crack. Record only when the sealant is in good condition for at least 90% of the length of the crack.

**FIGURE 19**
ACP 6. Low Severity
Transverse Cracking

**FIGURE 20**
ACP 6. Moderate Severity Transverse Cracking

**FIGURE 21**
ACP 6. High Severity Transverse Cracking

Cracking

This section includes the following distresses:

7. Patch/Patch Deterioration
8. Potholes

**Patching**

**and**

**Potholes**

# PATCH/PATCH DETERIORATION

## Description

Portion of pavement surface, greater than 0.1 sq. m (1 sq. ft), that has been removed and replaced or additional material applied to the pavement after original construction.

## Severity Levels

### LOW
Patch has at most low severity distress of any type.

### MODERATE
Patch has moderate severity distress of any type.

### HIGH
Patch has high severity distress of any type.

## How to Measure

Record number of patches and square meters (square feet) of affected surface area at each severity level.

Note: Any distress in the boundary of the patch is included in rating the patch.

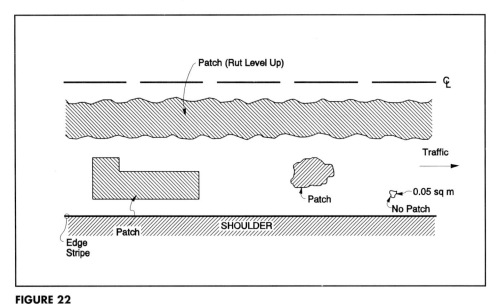

**FIGURE 22**
**ACP 7. Patch/Patch Deterioration**

**FIGURE 23**
ACP 7. Low Severity Patch

**FIGURE 24**
ACP 7. Moderate Severity Patch

**FIGURE 25**
ACP 7. High Severity Patch

Patching

and

Potholes

## POTHOLES

### Description

Bowl-shaped holes of various sizes in the pavement surface. Minimum plan dimension is 15 cm (6 in.).

### Severity Levels

**LOW**
Less than 25 mm (1 in.) deep

**MODERATE**
25 mm to 50 mm (1 to 2 in.) deep

**HIGH**
More than 50 mm (2 in.) deep

### How to Measure

Record number of potholes and square meters (square feet) of affected area at each severity level. Pothole depth is the maximum depth below pavement surface.

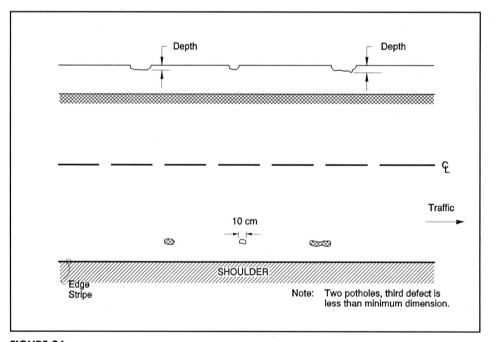

**FIGURE 26**
ACP 8. Potholes

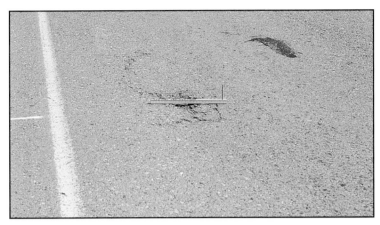

**FIGURE 27**
ACP 8. Low Severity Pothole

**FIGURE 28**
ACP 8. Moderate Severity Pothole

**FIGURE 29**
ACP 8. Moderate Severity Pothole—Close-up View

**FIGURE 30**
ACP 8. High Severity Pothole—Close-up View

This section includes the following types of surface deformations:

**9.** Rutting
**10.** Shoving

**Surface**

**Deformation**

# RUTTING

## Description

A rut is a longitudinal surface depression in the wheel path. It may have associated transverse displacement.

## Severity Levels

Not applicable. Severity levels could be defined by categorizing the measurements taken. A record of the measurements taken is much more desirable, however, because it is more accurate and repeatable than are severity levels.

## How to Measure

SPS-3 ONLY: Record maximum rut depth in millimeters, to the nearest millimeter, at 15-m (50-ft) intervals for each wheel path, as measured with a 1.2-m (4-ft) straight edge.

All other LTPP sections: Transverse profile is measured with a Dipstick profiler at 15-m (50-ft) intervals when not measured in conjunction with photographic distress surveys.

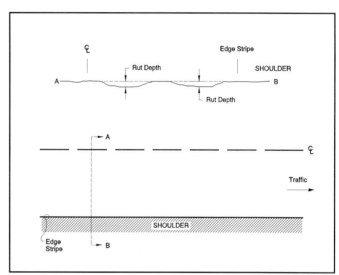

**FIGURE 31**
**ACP 9. Rutting**

**FIGURE 32**
**ACP 9. Rutting**

**FIGURE 33**
**ACP 9. Standing Water in Ruts**

**ASPHALT CONCRETE SURFACES**

# SHOVING

## Description

Shoving is a longitudinal displacement of a localized area of the pavement surface. It is generally caused by braking or accelerating vehicles, and is usually located on hills or curves, or at intersections. It also may have associated vertical displacement.

## Severity Levels

Not applicable. However, severity levels can be defined by the relative effect of shoving on ride quality.

## How to Measure

Record number of occurrences and square meters (square feet) of affected surface area.

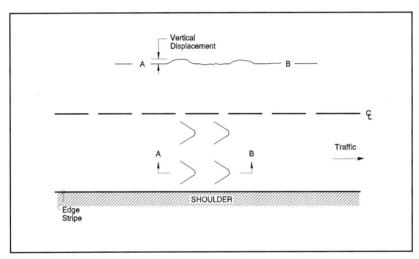

**FIGURE 34**
**ACP 10. Shoving**

**FIGURE 35**
**ACP 10. Shoving in Pavement Surface**

Surface

Deformation

This section includes the following types of surface defects:

**11.** Bleeding
**12.** Polished Aggregate
**13.** Raveling

**Surface**

**Defects**

# BLEEDING

### Description

Excess bituminous binder occurring on the pavement surface. May create a shiny, glass-like, reflective surface that may be tacky to the touch. Usually found in the wheel paths.

### Severity Levels

**LOW**

An area of pavement surface discolored relative to the remainder of the pavement by excess asphalt.

**MODERATE**

An area of pavement surface that is losing surface texture due to excess asphalt.

**HIGH**

Excess asphalt gives the pavement surface a shiny appearance; the aggregate may be obscured by excess asphalt; tire marks may be evident in warm weather.

### How to Measure

Record square meters (square feet) of surface area at each severity level.

**FIGURE 36**
ACP 11. Low Severity Bleeding

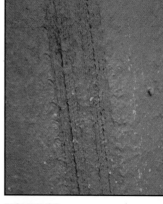

**FIGURE 38**
ACP 11. Tire Marks Evident in
High Severity Bleeding

**FIGURE 37**
ACP 11. Moderate Severity Bleeding

**ASPHALT
CONCRETE
SURFACES**

## POLISHED AGGREGATE

### Description

Surface binder worn away to expose coarse aggregate.

### Severity Levels

Not applicable.  However, the degree of polishing may be reflected in a reduction of surface friction.

### How to Measure

Record square meters (square feet) of affected surface area.

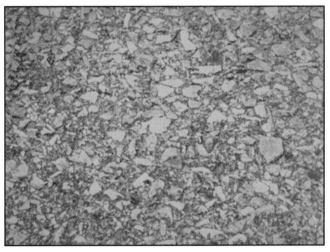

**FIGURE 39**
**ACP 12. Polished Aggregate**

# RAVELING

### Description

Wearing away of the pavement surface in high-quality hot mix asphalt concrete. Caused by the dislodging of aggregate particles and loss of asphalt binder.

### Severity Levels

**LOW**
The aggregate or binder has begun to wear away but has not progressed significantly. Some loss of fine aggregate.

**MODERATE**
Aggregate and/or binder has worn away and the surface texture is becoming rough and pitted; loose particles generally exist; loss of fine aggregate and some loss of coarse aggregate.

**HIGH**
Aggregate and/or binder has worn away and the surface texture is very rough and pitted; loss of coarse aggregate.

### How to Measure

Record square meters (square feet) of affected surface area at each severity level.

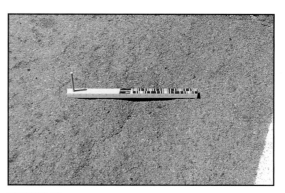

**FIGURE 40**
ACP 13. Low Severity Raveling

**FIGURE 41**
ACP 13. Moderate Severity Raveling

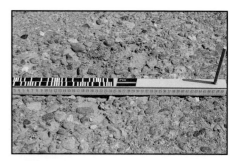

**FIGURE 42**
ACP 13. High Severity Raveling

This section includes the following distresses:

**14.** Lane-to-Shoulder Dropoff
**15.** Water Bleeding and Pumping

**Miscellaneous**

**Distresses**

**14**

## LANE-TO-SHOULDER DROPOFF

### Description

Difference in elevation between the traveled surface and the outside shoulder. Typically occurs when the outside shoulder settles as a result of pavement layer material differences.

### Severity Level

Not applicable. Severity levels could be defined by categorizing the measurements taken. A record of the measurements taken is much more desirable, however, because it is more accurate and repeatable than are severity levels.

### How to Measure

Record in millimeters (inches) to the nearest millimeter (0.04 in.), at intervals of 15 m (50 ft) along the lane-to-shoulder joint.

If the travelled surface is lower than the shoulder, record as a negative (-) value.

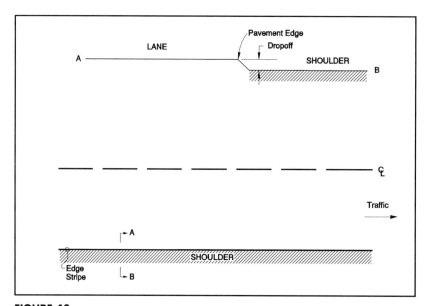

**FIGURE 43**
**ACP 14. Lane-to-Shoulder Dropoff**

**FIGURE 44**
**ACP 14. Lane-to-Shoulder Dropoff**

ASPHALT
CONCRETE
SURFACES

34

# WATER BLEEDING AND PUMPING

## Description

Seeping or ejection of water from beneath the pavement through cracks.

In some cases, detectable by deposits of fine material left on the pavement surface which were eroded (pumped) from the support layers and have stained the surface.

## Severity Levels

Not applicable. Severity levels are not used because the amount and degree of water bleeding and pumping changes with varying moisture conditions.

## How to Measure

Record the number of occurrences of water bleeding and pumping and the length in meters (feet) of affected pavement.

**FIGURE 45**
ACP 15. Water Bleeding and Pumping

**FIGURE 46**
ACP 15. Fine Material Left on Surface by Water
Bleeding and Pumping

Miscellaneous

Distresses

This section covers jointed (plain and reinforced) portland cement concrete-surfaced pavements (JCP), including jointed concrete overlays on portland cement concrete pavements. Each of the distresses has been grouped into one of the following categories:

    **A.** Cracking
    **B.** Joint Deficiencies
    **C.** Surface Defects
    **D.** Miscellaneous Distresses

Table 2 summarizes the various types of distress and unit of measurement. Some distresses also have defined severity levels.

| TABLE 2. Jointed Concrete-Surfaced Pavement Distress Types | | |
|---|---|---|
| **DISTRESS TYPE** | **UNIT OF MEASURE** | **DEFINED SEVERITY LEVELS?** |
| **A.** Cracking / page **39** | | |
|   1. Corner Breaks | Number | Yes |
|   2. Durability Cracking ("D" Cracking) | Number of Slabs, Square Meters | Yes |
|   3. Longitudinal Cracking | Meters | Yes |
|   4. Transverse Cracking | Number, Meters | Yes |
| **B.** Joint Deficiencies / page **47** | | |
|   5a. Transverse Joint Seal Damage | Number | Yes |
|   5b. Longitudinal Joint Seal Damage | Number, Meters | No |
|   6. Spalling of Longitudinal Joints | Meters | Yes |
|   7. Spalling of Transverse Joints | Number, Meters | Yes |
| **C.** Surface Defects / page **51** | | |
|   8a. Map Cracking | Number, Square Meters | No |
|   8b. Scaling | Number, Square Meters | No |
|   9. Polished Aggregate | Square Meters | No |
|   10. Popouts | Number/Square Meter | No |
| **D.** Miscellaneous Distress / page **55** | | |
|   11. Blowups | Number | No |
|   12. Faulting of Transverse Joints and Cracks | Millimeters | No |
|   13. Lane-to-Shoulder Dropoff | Millimeters | No |
|   14. Lane-to-Shoulder Separation | Millimeters | No |
|   15. Patch/Patch Deterioration | Number, Square Meters | Yes |
|   16. Water Bleeding and Pumping | Number, Meters | No |

**2**

**DISTRESSES FOR PAVEMENTS WITH JOINTED PORTLAND CEMENT CONCRETE SURFACES**

This section includes the following types of distresses:

1. Corner Breaks
2. Durability Cracking ("D" Cracking)
3. Longitudinal Cracking
4. Transverse Cracking

Figure 47 illustrates the proper measurement of crack width and width of spalling for cracks and joints.

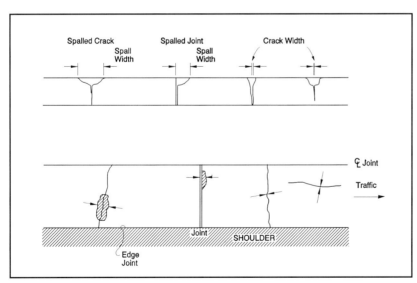

**FIGURE 47**
**Measuring Widths of Spalls and Cracks in Jointed Concrete Pavement**

## CORNER BREAKS

### Description

A portion of the slab separated by a crack which intersects the adjacent transverse and longitudinal joints, describing approximately a 45° angle with the direction of traffic. The length of the sides is from 0.3 m (1 ft) to one-half the width of the slab, on each side of the corner.

### Severity Levels

**LOW**
Crack is not spalled for more than 10% of the length of the crack; there is no measurable faulting; and the corner piece is not broken into two or more pieces.

**MODERATE**
Crack is spalled at low severity for more than 10% of its total length; or faulting of crack or joint is < 13 mm (0.5 in.); and the corner piece is not broken into two or more pieces.

**HIGH**
Crack is spalled at moderate to high severity for more than 10% of its total length; or faulting of the crack or joint is ≥ 13 mm (0.5 in.); or the corner piece is broken into two or more pieces.

### How to Measure

Record number of corner breaks at each severity level.

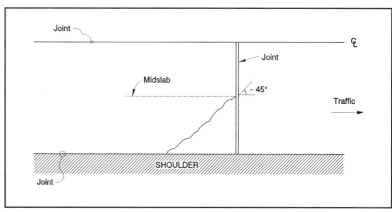

**FIGURE 48**
**JCP 1. Corner Breaks**

**FIGURE 49**
**JCP 1. Low Severity Corner Break**

**FIGURE 50**
**JCP 1. Moderate Severity Corner Break**

# DURABILITY CRACKING ("D" CRACKING)

## Description

Closely spaced crescent-shaped hairline cracking pattern.

Occurs adjacent to joints, cracks, or free edges; initiating in slab corners.

Dark coloring of the cracking pattern and surrounding area.

## Severity Levels

### LOW

"D" cracks are tight, with no loose or missing pieces, and no patching is in the affected area.

### MODERATE

"D" cracks are well defined, and some small pieces are loose or have been displaced.

### HIGH

"D" cracking has a well-developed pattern, with a significant amount of loose or missing material. Displaced pieces, up to 0.1 sq. m (1 sq. ft), may have been patched.

## How to Measure

Record number of slabs with "D" cracking and square meters (square feet) of area affected at each severity level. The slab severity rating is based on the highest severity level present for at least 10% of the affected area.

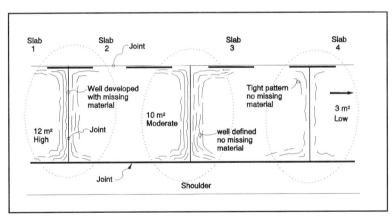

**FIGURE 51**
**JCP 2. Durability Cracking ("D" Cracking)**

**FIGURE 53**
**JCP 2. High Severity "D" Cracking with Loose and Missing Material**

**FIGURE 52**
**JCP 2. Moderate Severity "D" Cracking with Well-Defined Pattern**

Cracking

# LONGITUDINAL CRACKING

## Description

Cracks that are predominantly parallel to the pavement centerline

## Severity Levels

### LOW
Crack widths < 3 mm (0.125 in.), no spalling, and no measurable faulting; or well sealed and with a width that cannot be determined.

### MODERATE
Crack widths ≥ 3 mm (0.125 in.) and < 13 mm (0.5 in.); or with spalling < 75 mm (3 in.); or faulting up to 13 mm (0.5 in.).

### HIGH
Crack widths ≥ 13 mm (0.5 in.); or with spalling ≥ 75 mm (3 in.); or faulting ≥ 13 mm (0.5 in.).

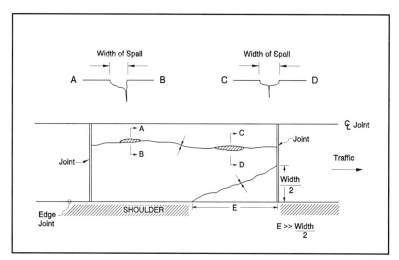

**FIGURE 54**
**JCP 3. Longitudinal Cracking**

**How to Measure**

Record length in meters (feet) of longitudinal cracking at each severity level.

Also record length in meters (feet) of longitudinal cracking with sealant in good condition at each severity level.

**FIGURE 55**
JCP 3. Low Severity Longitudinal Cracking

**FIGURE 56**
JCP 3. Moderate Severity Longitudinal Cracking

**FIGURE 57**
JCP 3. High Severity Longitudinal Cracking

# 4

# TRANSVERSE CRACKING

### Description

Cracks that are predominantly perpendicular to the pavement centerline

### Severity Levels

#### LOW
Crack widths < 3 mm (0.125 in.), no spalling, and no measurable faulting; or well-sealed and the width cannot be determined.

#### MODERATE
Crack widths ≥ 3 mm (0.125 in.) and < 6 mm (0.25 in.); or with spalling < 75 mm (3 in.); or faulting up to 6 mm (0.25 in.).

#### HIGH
Crack widths ≥ 6 mm (0.25 in.); or with spalling ≥ 75 mm (3 in.); or faulting ≥ 6 mm (0.25 in.).

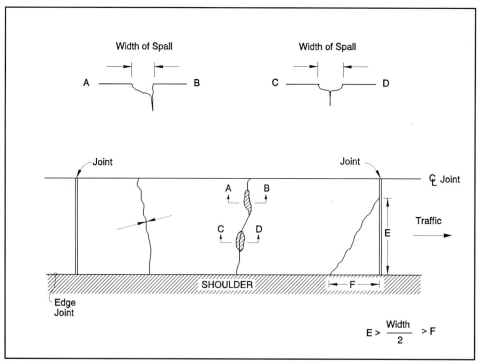

**FIGURE 58**
JCP 4. Transverse Cracking

**How to Measure**

Record number and length of transverse cracks at each severity level. Rate the entire transverse crack at the highest severity level present for at least 10% of the total length of the crack. Length recorded, in meters (feet), is the total length of the crack and is assigned to the highest severity level present for at least 10% of the total length of the crack.

Also record the length, in meters (feet), of transverse cracking at each severity level with sealant in good condition. The length recorded, in meters (feet), is the total length of the well-sealed crack and is assigned to the severity level of the crack. Record only when the sealant is in good condition for at least 90% of the length of the crack.

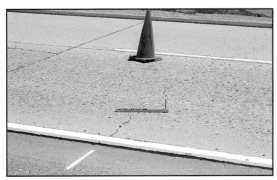

**FIGURE 59**
**JCP 4. Moderate Severity Transverse Cracking**

**FIGURE 60**
**JCP 4. High Severity Transverse Cracking**

This section includes the following types of distresses:

**5a.** Transverse Joint Seal Damage
**5b.** Longitudinal Joint Seal Damage
**6.** Spalling of Longitudinal Joints
**7.** Spalling of Transverse Joints

# B

# Joint
# Deficiencies

# JOINT SEAL DAMAGE

## Description

Joint seal damage is any condition which enables incompressible materials or a significant amount of water to infiltrate the joint from the surface. Typical types of joint seal damage are:

Extrusion, hardening, adhesive failure (bonding), cohesive failure (splitting), or complete loss of sealant.

Intrusion of foreign material in the joint.

Weed growth in the joint.

## 5a. TRANSVERSE JOINT SEAL DAMAGE

### Severity Levels

**LOW**
Joint seal damage as described above exists over less than 10% of the joint.

**MODERATE**
Joint seal damage as described above exists over 10-50% of the joint.

**HIGH**
Joint seal damage as described above exists over more than 50% of the joint.

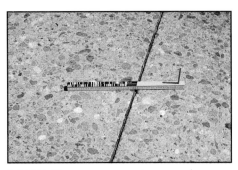

**FIGURE 61**
JCP 5. Low Severity Joint Seal Damage

### How to Measure

Indicate whether the transverse joints have been sealed (yes or no).
If yes, record number of sealed transverse joints at each severity level.

## 5b. LONGITUDINAL JOINT SEAL DAMAGE

### Severity Levels

None

### How to Measure

Record number of longitudinal joints that are sealed (0, 1, 2).

Record total length of sealed longitudinal joints with joint seal damage as described above. Individual occurrences are recorded only when at least 1 m (3.3 ft) in length.

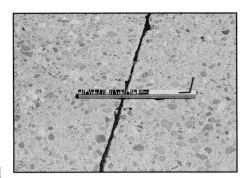

**FIGURE 62**
JCP 5. Moderate Severity Joint Seal Damage

# SPALLING OF LONGITUDINAL JOINTS

## Description

Cracking, breaking, chipping, or fraying of slab edges within 0.6 m (2 ft) of the longitudinal joint.

## Severity Levels

### LOW
Spalls less than 75 mm (3 in.) wide, measured to the center of the joint, with loss of material, or spalls with no loss of material and no patching.

### MODERATE
Spalls 75 mm (3 in.) to 150 mm (6 in.) wide, measured to the center of the joint, with loss of material.

### HIGH
Spalls greater than 150 mm (6 in.) wide, measured to the center of the joint, with loss of material.

## How to Measure

Record length in meters (feet) of longitudinal joint spalling at each severity level.

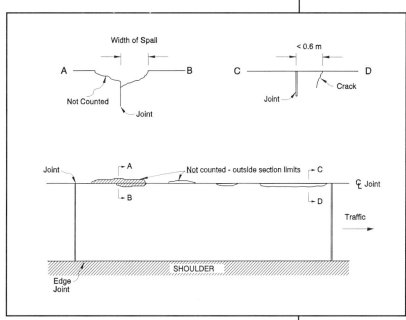

**FIGURE 63**
JCP 6. Spalling of Longitudinal Joints

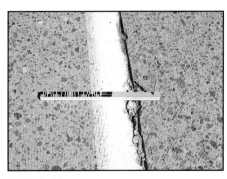

**FIGURE 64**
JCP 6. Low Severity Spalling of Longitudinal Joint

**FIGURE 65**
JCP 6. High Severity Spalling of Longitudinal Joint

Joint
Deficiencies

# 7

# SPALLING OF TRANSVERSE JOINTS

### Description

Cracking, breaking, chipping, or fraying of slab edges within 0.6 m (2 ft) of transverse joint.

### Severity Levels

#### LOW

Spalls less than 75 mm (3 in.) wide, measured to the center of the joint, with loss of material, or spalls with no loss of material and no patching.

#### MODERATE

Spalls 75 mm (3 in.) to 150 mm (6 in.) wide, measured to the center of the joint, with loss of material.

#### HIGH

Spalls greater than 150 mm (6 in.) wide, measured to the center of the joint, with loss of material.

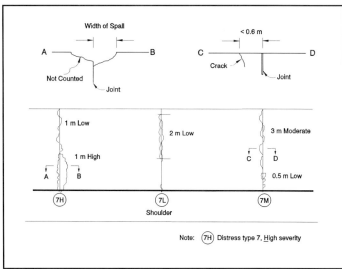

**FIGURE 66**
JCP 7. Spalling of Transverse Joints

### How to Measure

Record number of affected transverse joints at each severity level. Rate the entire transverse joint at the highest severity level present for at least 10% of the total length of the joint.

Record length in meters (feet) of the spalled portion of the joint. Record at the highest severity level present for at least 10% of the total length of the spalling.

**FIGURE 67**
JCP 7. Moderate Severity Spalling of Transverse Joint—Far View

**FIGURE 68**
JCP 7. Moderate Severity Spalling of Transverse Joint—Close-up View

This section includes the following types of distresses:

**8a.** Map Cracking
**8b.** Scaling
**9.** Polished Aggregate
**10.** Popouts

C

**Surface**
**Defects**

# MAP CRACKING AND SCALING

## 8a. *MAP CRACKING*

### Description

A series of cracks that extend only into the upper surface of the slab. Frequently, larger cracks are oriented in the longitudinal direction of the pavement and are interconnected by finer transverse or random cracks.

### Severity Levels

Not applicable

### How to Measure

Record the number of occurrences and the square meters (square feet) of affected area.

## 8b. *SCALING*

### Description

Scaling is the deterioration of the upper concrete slab surface, normally 3 mm (0.125 in.) to 13 mm (0.5 in.), and may occur anywhere over the pavement.

### Severity Levels

Not applicable

### How to Measure

Record the number of occurrences and the square meters (square feet) of affected area.

**FIGURE 69**
JCP 8a. Map Cracking

**FIGURE 70**
JCP 8b. Scaling

**FIGURE 71**
JCP 8b. Scaling—Close-up View

## POLISHED AGGREGATE

### Description

Surface mortar and texturing worn away to expose coarse aggregate.

### Severity Levels

Not applicable. However, the degree of polishing may be reflected in a reduction of surface friction.

### How to Measure

Record square meters (square feet) of affected surface area.

**FIGURE 72**
**JCP 9. Polished Aggregate**

## POPOUTS

### Description

Small pieces of pavement broken loose from the surface, normally ranging in diameter from 25 mm (1 in.) to 100 mm (4 in.) and depth from 13 mm (0.5 in.) to 50 mm (2 in.).

### Severity Levels

Not applicable. However, severity levels can be defined in relation to the intensity of popouts as measured below.

### How to Measure

Record number of popouts per square meter (square foot).

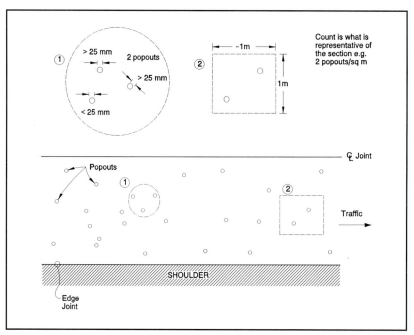

**FIGURE 73**
**JCP 10. Popouts**

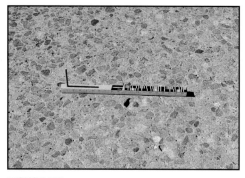

**FIGURE 74**
**JCP 10. A Popout**

This section includes the following distresses:

**11.** Blowups
**12.** Faulting of Transverse Joints and Cracks
**13.** Lane-to-Shoulder Dropoff
**14.** Lane-to-Shoulder Separation
**15.** Patch/Patch Deterioration
**16.** Water Bleeding and Pumping

**Miscellaneous**

**Distresses**

# BLOWUPS

### Description

Localized upward movement of the pavement surface at transverse joints or cracks, often accompanied by shattering of the concrete in that area.

### Severity Levels

Not applicable.  However, severity levels can be defined by the relative effect of a blowup on ride quality and safety.

### How to Measure

Record the number of blowups.

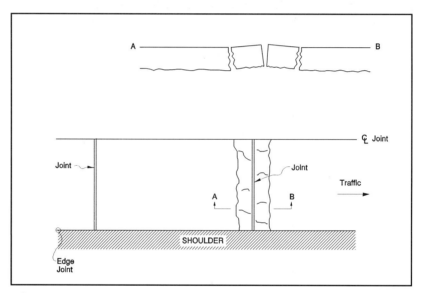

**FIGURE 75**
**JCP 11. Blowups**

**FIGURE 76**
**JCP 11. A Blowup**

# FAULTING OF TRANSVERSE JOINTS AND CRACKS

## Description

Difference in elevation across a joint or crack.

## Severity Levels

Not applicable. Severity levels could be defined by categorizing the measurements taken. A complete record of the measurements taken is much more desirable, however, because it is more accurate and repeatable than are severity levels.

## How to Measure

Record in millimeters (inches) to the nearest millimeter (0.04 in.); 0.3 m (1 ft) from the outside lane edge and 0.75 m (2.5 ft) from the outside lane edge (wheel path).

If the "approach" slab is higher than the "departure" slab, record faulting as positive (+); if the approach slab is lower, record faulting as negative (-). (See Appendix C.)

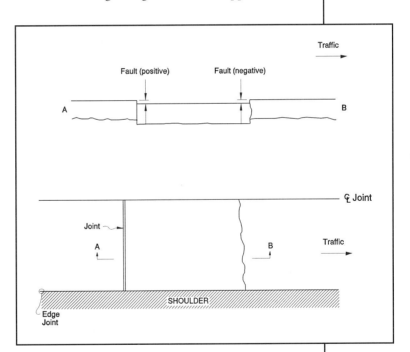

**FIGURE 77**
JCP 12. Faulting of Transverse Joints and Cracks

**FIGURE 78**
JCP 12. Faulting of Transverse Cracks

## LANE-TO-SHOULDER DROPOFF

### Description

Difference in elevation between the edge of slab and outside shoulder; typically occurs when the outside shoulder settles.

### Severity Levels

Not applicable.  Severity levels could be defined by categorizing the measurements taken.  A complete record of the measurements taken is much more desirable, however, because it is more accurate and repeatable than are severity levels.

### How to Measure

Measure at the longitudinal construction joint between the lane edge and the shoulder.

Record in millimeters (inches) to the nearest millimeter (0.04 in.) at 15-m (50-ft) intervals along the lane-to-shoulder joint.

If the travelled surface is lower than the shoulder, record as a negative (-) value.

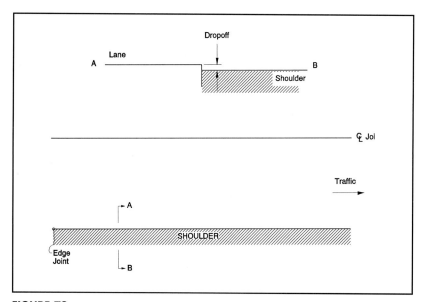

**FIGURE 79**
**JCP 13. Lane-to-Shoulder Dropoff**

**FIGURE 80**
**JCP 13. Lane-to-Shoulder Dropoff**

# LANE-TO-SHOULDER SEPARATION

## Description

Widening of the joint between the edge of the slab and the shoulder.

## Severity Levels

Not applicable. Severity levels could be defined by categorizing the measurements taken. A complete record of the measurements taken is much more desirable, however, because it is more accurate and repeatable than are severity levels.

## How to Measure

Record in millimeters (inches) to the nearest millimeter (0.04 in.), at intervals of 15 m (50 ft) along the lane-to-shoulder joint. Indicate whether the joint is well sealed (yes or no) at each location.

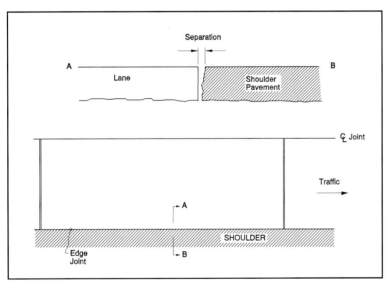

**FIGURE 81**
JCP 14. Lane-to-Shoulder Separation

**FIGURE 82**
JCP 14. Poorly Sealed
Lane-to-Shoulder Separation

**FIGURE 83**
JCP 14. Well-Sealed
Lane-to-Shoulder Separation

# 15

# PATCH/PATCH DETERIORATION

## Description

A portion, greater than 0.1 sq. m (1 sq. ft), or all of the original concrete slab that has been removed and replaced, or additional material applied to the pavement after original construction.

## Severity Levels

### LOW

Patch has at most low severity distress of any type; and no measurable faulting or settlement at the perimeter of the patch.

### MODERATE

Patch has moderate severity distress of any type; or faulting or settlement up to 6 mm (0.25 in.) at the perimeter of the patch.

### HIGH

Patch has a high severity distress of any type; or faulting or settlement ≥ 6 mm (0.25 in.) at the perimeter of the patch.

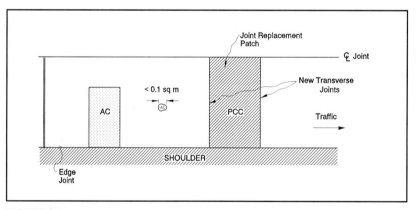

**FIGURE 84**
**JCP 15. Patch/Patch Deterioration**

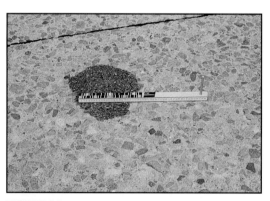

**FIGURE 85**
**JCP 15. Small, Low Severity Asphalt Concrete Patch**

## How to Measure

Record number of patches and square meters (square feet) of affected surface area at each severity level, recorded separately by material type—rigid versus flexible.

For slab replacement, rate each slab as a separate patch and continue to rate joints.

**FIGURE 86**
JCP 15. Large, Low Severity Asphalt Concrete Patch

**FIGURE 87**
JCP 15. Large, High Severity Asphalt Concrete Patch

**FIGURE 88**
JCP 15. Large, Low Severity Portland Cement Concrete Patch

# 16

## WATER BLEEDING AND PUMPING

### Description

Seeping or ejection of water from beneath the pavement through cracks.

In some cases detectable by deposits of fine material left on the pavement surface, which were eroded (pumped) from the support layers and have stained the surface.

### Severity Levels

Not applicable. Severity levels are not used because the amount and degree of water bleeding and pumping changes with varying moisture conditions.

### How to Measure

Record the number of occurrences of water bleeding and pumping and the length in meters (feet) of affected pavement.

**FIGURE 89**
**JCP 16. Water Bleeding and Pumping**

This section covers continuously reinforced concrete-surfaced pavements (CRCP), including continuously reinforced concrete overlays on portland cement concrete pavements. Each of the distresses has been grouped into one of the following categories:

A. Cracking
B. Surface Defects
C. Miscellaneous Distresses

Table 3 summarizes the various types of distress and unit of measurement. Some distresses also have defined severity levels.

| **TABLE 3.** Continuously Reinforced Concrete-Surfaced Pavement Distress Types | | |
|---|---|---|
| **DISTRESS TYPE** | **UNIT OF MEASURE** | **DEFINED SEVERITY LEVELS?** |
| **A.** Cracking / page **65** | | |
| 1. Durability Cracking ("D" Cracking) | Number, Square Meters | Yes |
| 2. Longitudinal Cracking | Meters | Yes |
| 3. Transverse Cracking | Number, Meters | Yes |
| **B.** Surface Defects / page **71** | | |
| 4a. Map Cracking | Number, Square Meters | No |
| 4b. Scaling | Number, Square Meters | No |
| 5. Polished Aggregate | Square Meters | No |
| 6. Popouts | Number/Square Meter | No |
| **C.** Miscellaneous Distresses / page **75** | | |
| 7. Blowups | Number | No |
| 8. Transverse Construction Joint Deterioration | Number | Yes |
| 9. Lane-to-Shoulder Dropoff | Millimeters | No |
| 10. Lane-to-Shoulder Separation | Millimeters | No |
| 11. Patch/Patch Deterioration | Number, Square Meters | Yes |
| 12. Punchouts | Number | Yes |
| 13. Spalling of Longitudinal Joints | Meters | Yes |
| 14. Water Bleeding and Pumping | Number, Meters | No |
| 15. Longitudinal Joint Seal Damage | Number, Meters | No |

This section includes the following distresses:

1. Durability Cracking ("D" Cracking)
2. Longitudinal Cracking
3. Transverse Cracking

# Cracking

# DURABILITY CRACKING ("D" CRACKING)

## Description

Closely spaced crescent-shaped hairline cracking pattern

Occurs adjacent to joints, cracks, or free edges. Initiates at the intersection, e.g., cracks and a free edge.

Dark coloring of the cracking pattern and surrounding area

## Severity Levels

### LOW
"D" cracks are tight, with no loose or missing pieces, and no patching is in the affected area.

### MODERATE
"D" cracks are well defined, and some small pieces are loose or have been displaced.

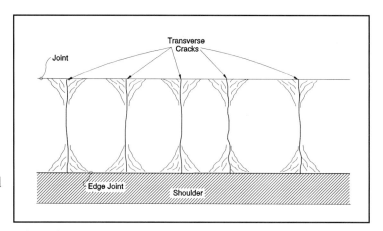

**FIGURE 90**
CRCP 1. Durability Cracking ("D" Cracking)

### HIGH
"D" cracking has a well-developed pattern, with a significant amount of loose or missing material. Displaced pieces, up to 0.1 sq. m (1 sq. ft), may have been patched.

## How to Measure

Record number of affected transverse cracks at each severity level and the square meters (square feet) of area affected at each severity level. The "D" cracking severity rating is based on the highest severity level present for at least 10% of the affected area.

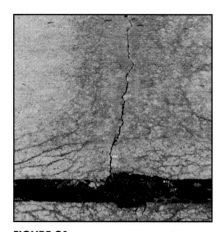

**FIGURE 91**
CRCP 1. Moderate Severity "D" Cracking at Transverse Crack

**FIGURE 92**
CRCP 1. High Severity "D" Cracking at Longitudinal Joint

**CONTINUOUSLY REINFORCED CONCRETE SURFACES**

# LONGITUDINAL CRACKING

## Description

Cracks that are predominantly parallel to the pavement centerline.

## Severity Levels

### LOW
Crack widths < 3 mm (0.125 in.), no spalling, and there is no measurable faulting; or well sealed and with a width that cannot be determined.

### MODERATE
Crack widths ≥ 3 mm (0.125 in.) and < 13 mm (0.5 in.); or with spalling < 75 mm (3 in.); or faulting up to 13 mm (0.5 in.).

### HIGH
Crack widths ≥ 13 mm (0.5 in.); or with spalling ≥ 75 mm (3 in.); or faulting ≥ 13 mm (0.5 in.).

## How to Measure

Record length in meters (feet) of longitudinal cracking at each severity level.

Also record length in meters (feet) of longitudinal cracking with sealant in good condition at each severity level.

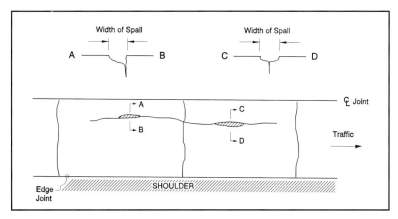

**FIGURE 93**
**CRCP 2. Longitudinal Cracking**

**FIGURE 94**
**CRCP 2. Low Severity Longitudinal Cracking**

**FIGURE 95**
**CRCP 2. High Severity Longitudinal Cracking**

Cracking

67

# 3

# TRANSVERSE CRACKING

### Description

Cracks that are predominantly perpendicular to the pavement centerline. This cracking is expected in a properly functioning continuously reinforced concrete pavement. "Y" cracks are routine, naturally occurring defects, and shall be counted as a single occurrence of a transverse crack.

### Severity Levels

**LOW**
Cracks that are spalled along ≤ 10% of the crack length.

**MODERATE**
Cracks that are spalled along > 10% and ≤ 50% of the crack length.

**HIGH**
Cracks that are spalled along > 50% of the crack length.

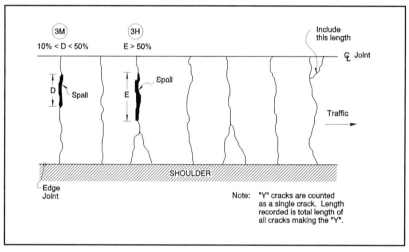

**FIGURE 96**
CRCP 3. Transverse Cracking

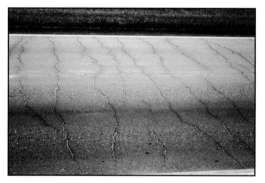

**FIGURE 97**
CRCP 3. Transverse Cracking Pattern

## How to Measure

Record the total number of transverse cracks within the survey section, including those that are not distressed.

Record separately the number and length in meters (feet) of transverse cracking at each severity level. Length recorded, in meters (feet), is the total length of the crack.

"Y" cracks shall be considered as single cracks. The sum of the individual crack lengths shall be recorded.

**FIGURE 98**
CRCP 3. Low Severity Transverse Cracking

**FIGURE 99**
CRCP 3. Moderate Severity Transverse Cracking

**FIGURE 100**
CRCP 3. High Severity Transverse Cracking

Cracking

69

This section includes the following:

B

**Surface**

**Defects**

# MAP CRACKING AND SCALING

### 4a. *MAP CRACKING*

## Description

A series of cracks that extend only into the upper surface of the slab. Frequently, larger cracks are oriented in the longitudinal direction of the pavement and are interconnected by finer transverse or random cracks.

## Severity Levels

Not applicable

## How to Measure

Record the number of occurrences and the square meters (square feet) of affected area.

### 4b. *SCALING*

## Description

Scaling is the deterioration of the upper concrete slab surface, normally 3 mm (0.125 in.) to 13 mm (0.5 in.), and may occur anywhere over the pavement.

## Severity Levels

Not applicable

## How to Measure

Record the number of occurrences and the square meters (square feet) of affected area.

**FIGURE 101**
**CRCP 4a. Map Cracking Attributable to Alkali-Silica Reactivity (ASR)**

**FIGURE 102**
**CRCP 4b. Scaling**

## POLISHED AGGREGATE

### Description

Surface mortar and texturing worn away to expose coarse aggregate.

### Severity Levels

Not applicable. However, the degree of polishing may be reflected in a reduction of surface friction.

### How to Measure

Record square meters (square feet) of affected surface area.

**FIGURE 103**
**CRCP 5. Polished Aggregate**

# POPOUTS

### Description

Small pieces of pavement broken loose from the surface, normally ranging in diameter from 25 mm (1 in.) to 100 mm (4 in.) and depth from 13 mm (0.5 in.) to 50 mm (2 in.).

### Severity Levels

Not applicable.  However, severity levels can be defined in relation to the intensity of popouts as measured below.

### How to Measure

Record number of popouts per square meter (square foot).

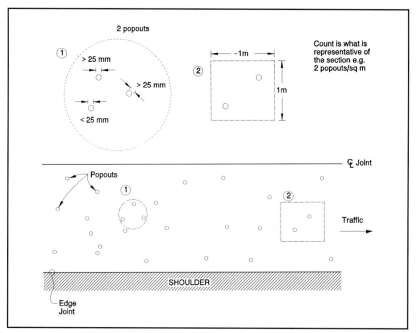

**FIGURE 104**
CRCP 6. Popouts

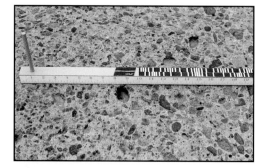

**FIGURE 105**
CRCP 6. Popouts

This section includes the following distresses:

C

**Miscellaneous**

**Distresses**

# BLOWUPS

### Description

Localized upward movement of the pavement surface at transverse joints or cracks, often accompanied by shattering of the concrete in that area.

### Severity Levels

Not applicable. However, severity levels can be defined by the relative effect of a blowup on ride quality and safety.

### How to Measure

Record number of blowups.

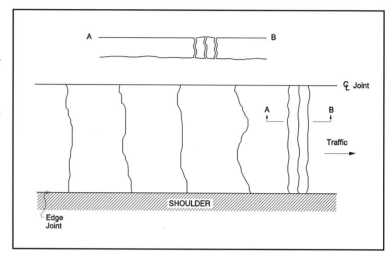

**FIGURE 106**
CRCP 7. Blowups

**FIGURE 107**
CRCP 7. A Blowup

**FIGURE 108**
CRCP 7. Close-Up View of a Blowup

**FIGURE 109**
CRCP 7. Exposed Steel in a Blowup

CONTINUOUSLY
REINFORCED
CONCRETE
SURFACES

# TRANSVERSE CONSTRUCTION JOINT DETERIORATION

## Description

Series of closely spaced transverse cracks or a large number of interconnecting cracks occurring near the construction joint.

## Severity Levels

### LOW
No spalling or faulting within 0.6 m (2 ft) of construction joint.

### MODERATE
Spalling < 75 mm (3 in.) exists within 0.6 m (2 ft) of construction joint.

### HIGH
Spalling ≥ 75 mm (3 in.) and breakup exists within 0.6 m (2 ft) of construction joint.

## How to Measure

Record number of construction joints at each severity level.

**FIGURE 110**
CRCP 8. Transverse Construction Joint Deterioration

**FIGURE 111**
CRCP 8. Low Severity Transverse Construction Joint Deterioration

**FIGURE 112**
CRCP 8. Moderate Severity Transverse Construction Joint Deterioration

**FIGURE 113**
CRCP 8. Attempted Repair of High Severity Transverse Construction Joint Deterioration

Miscellaneous

Distresses

## LANE-TO-SHOULDER DROPOFF

### Description

Difference in elevation between the edge of slab and outside shoulder; typically occurs when the outside shoulder settles.

### Severity Levels

Not applicable. Severity levels could be defined by categorizing the measurements taken. A complete record of the measurements taken is much more desirable, however, because it is more accurate and repeatable than are severity levels.

### How to Measure

Measure at the longitudinal construction joint between the lane edge and the shoulder.

Record in millimeters (inches) to the nearest millimeter (0.04 in.) at 15-m (50-ft) intervals along the lane-to-shoulder joint.

If the travelled surface is lower than the shoulder, record as a negative (-) value.

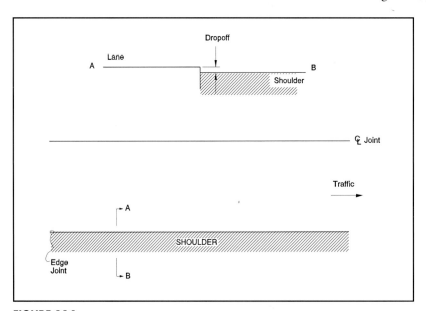

**FIGURE 114**
CRCP 9. Lane-to-Shoulder Dropoff

**FIGURE 115**
CRCP 9. Lane-to-Shoulder Dropoff

# LANE-TO-SHOULDER SEPARATION

### Description

Widening of the joint between the edge of the slab and the shoulder.

### Severity Levels

Not applicable.  Severity levels could be defined by categorizing the measurements taken.  A complete record of the measurements taken is much more desirable, however, because it is more accurate and repeatable than are severity levels.

### How to Measure

Record in millimeters (inches) to the nearest millimeter (0.04 in.), at intervals of 15 m (50 ft) along the lane-to-shoulder joint, and indicate whether the joint is well sealed (yes or no) at each location.

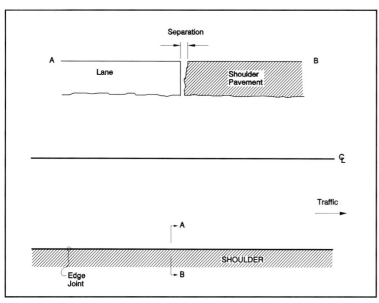

**FIGURE 116**
CRCP 10. Lane-to-Shoulder Separation

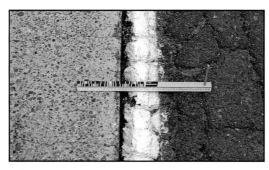

**FIGURE 117**
CRCP 10. Close-Up View of a Lane-to-Shoulder Separation

Miscellaneous

Distresses

# 11

# PATCH/PATCH DETERIORATION

## Description

A portion, greater than 0.1 sq. m (1 sq. ft), or all of the original concrete slab that has been removed and replaced, or additional material applied to the pavement after original construction.

## Severity Levels

### LOW
Patch has at most low severity distress of any type; and no measurable faulting or settlement at the perimeter of the patch.

### MODERATE
Patch has moderate severity distress of any type; or faulting or settlement up to 6 mm (0.25 in.) at the perimeter of the patch.

### HIGH
Patch has a high severity distress of any type; or faulting or settlement ≥ 6 mm (0.25 in.) at the perimeter of the patch.

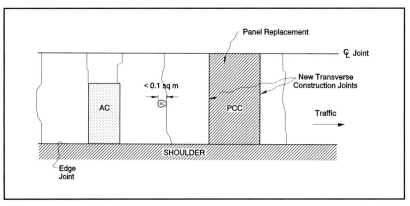

**FIGURE 118**
**CRCP 11. Patch/Patch Deterioration**

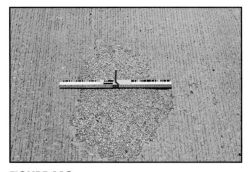

**FIGURE 119**
**CRCP 11. Small, Low Severity Asphalt Concrete Patch**

**How to Measure**

Record number of patches and square meters (square feet) of affected surface area at each severity level, recorded separately by material type—rigid versus flexible.

Note: Panel replacement shall be rated as a patch. New transverse cracks shall be rated separately. Any sawn joints shall be considered construction joints and rated separately.

**FIGURE 120**
CRCP 11. Low Severity Asphalt Concrete Patch

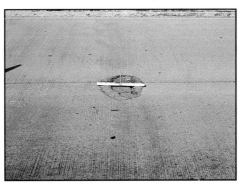

**FIGURE 121**
CRCP 11. Moderate Severity Asphalt Concrete Patch

**FIGURE 122**
CRCP 11. Low Severity Portland Cement Concrete Patch

# 12

# PUNCHOUTS

## Description

The area enclosed by two closely spaced (usually less than 0.6 m [2 ft]) transverse cracks, a short longitudinal crack, and the edge of the pavement or a longitudinal joint. Also includes "Y" cracks that exhibit spalling, breakup, and faulting.

## Severity Levels

### LOW
Longitudinal and transverse cracks are tight; and may have spalling < 75 mm (3 in.) or faulting < 6 mm (0.25 in.). Does not include "Y" cracks.

### MODERATE
Spalling ≥ 75 mm (3 in.) and < 150 mm (6 in.) or faulting ≥ 6 mm (0.25 in.) and < 13 mm (0.5 in.) exists.

### HIGH
Spalling ≥ 150 mm (6 in.) or concrete within the punchout is punched down by ≥ 13 mm (0.5 in.) or is loose and moves under traffic.

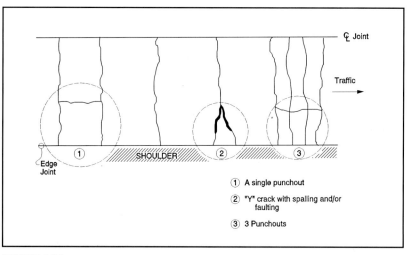

**FIGURE 123**
**CRCP 12. Punchouts**

**FIGURE 124**
**CRCP 12. Low Severity Punchout**

**How to Measure**

Record number of punchouts at each severity level.

The cracks which outline the punchout are also recorded under "Longitudinal Cracking" (CRCP 2) and "Transverse Cracking" (CRCP 3) .

**FIGURE 125**
CRCP 12. Moderate Severity Punchout

**FIGURE 126**
CRCP 12. High Severity Punchout

## SPALLING OF LONGITUDINAL JOINTS

### Description

Cracking, breaking, chipping, or fraying of slab edges within 0.6 m (2 ft) of the longitudinal joint.

### Severity Levels

#### LOW
Spalls less than 75 mm (3 in.) wide, measured to the center of the joint, with loss of material or spalls with no loss of material and no patching.

#### MODERATE
Spalls 75 mm (3 in.) to 150 mm (6 in.) wide, measured to the center of the joint, with loss of material.

#### HIGH
Spalls greater than 150 mm (6 in.) wide, measured to the center of the joint, with loss of material.

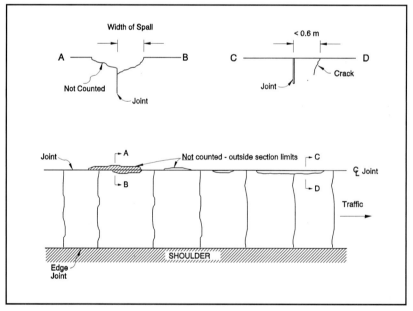

**FIGURE 127**
CRCP 13. Spalling of Longitudinal Joints

## How to Measure

Record length in meters (feet) of longitudinal joint spalling at each severity level.

**FIGURE 128**
CRCP 13. Close-Up View of Low Severity
Spalling of a Longitudinal Joint

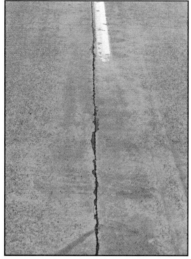

**FIGURE 129**
CRCP 13. Low Severity Spalling
of a Longitudinal Joint

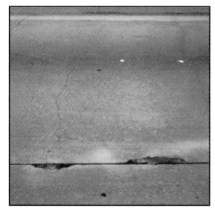

**FIGURE 130**
CRCP 13. Moderate Severity Spalling of a Longitudinal Joint

# WATER BLEEDING AND PUMPING

### Description

Seeping or ejection of water from beneath the pavement through cracks or joints.

In some cases detectable by deposits of fine material left on the pavement surface, which were eroded (pumped) from the support layers and have stained the surface.

### Severity Levels

Not applicable. Severity levels are not used because the amount and degree of water bleeding and pumping changes with varying moisture conditions.

### How to Measure

Record the number of occurrences of water bleeding and pumping and the length in meters (feet) of affected pavement.

**FIGURE 131**
CRCP 14. Water Bleeding and Pumping

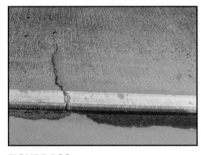

**FIGURE 132**
CRCP 14. Close-Up View of Water Bleeding and Pumping

# LONGITUDINAL JOINT SEAL DAMAGE

### Description

Joint seal damage is any condition which enables incompressible materials or a significant amount of water to infiltrate into the joint from the surface. Typical types of joint seal damage are:

Extrusion, hardening, adhesive failure (bonding), cohesive failure (splitting), or complete loss of sealant.

Intrusion of foreign material in the joint.

Weed growth in the joint.

### Severity Levels

Not applicable.

### How to Measure

Record number of longitudinal joints that are sealed (0, 1, 2).

Record length of sealed longitudinal joints with joint seal damage as described above.

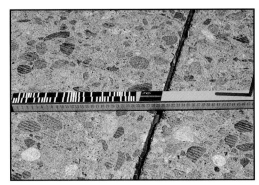

**FIGURE 133**
**CRCP 15. Longitudinal Joint Seal Damage**

## Glossary

**ADHESIVE FAILURE**
loss of bond (i.e. between the joint sealant and the joint reservoir; between the aggregate and the binder)

**AGGREGATE INTERLOCK**
interaction of aggregate particles across cracks and joints to transfer load

**APPROACH SLAB**
section of pavement just prior to joint, crack, or other significant roadway feature relative to the direction of traffic (see also leave slab)

**BINDER**
brown or black adhesive material used to hold stones together for paving

**BITUMINOUS**
like or from asphalt

**BLEEDING**
identified by a film of bituminous material on the pavement surface that creates a shiny, glass-like, reflective surface that may be tacky to the touch in warm weather

**BLOCK CRACKING**
the occurrence of cracks that divide the asphalt surface into approximately rectangular pieces, typically one square foot or more in size

**BLOWUP**
the result of localized upward movement or shattering of a slab along a transverse joint or crack

**CENTERLINE**
the painted line separating traffic lanes

**CHIPPING**
breaking or cutting off small pieces from the surface

**COHESIVE FAILURE**
the loss of a material's ability to bond to itself. Results in the material splitting or tearing apart from itself (i.e., joint sealant splitting)

**CONSTRUCTION JOINT**
the point at which work is concluded and reinitiated when building a pavement

**CORNER BREAK**
a portion of a jointed concrete pavement separated from the slab by a diagonal crack intersecting the transverse and longitudinal joint, which extends down through the slab, allowing the corner to move independently from the rest of the slab

**DURABILITY CRACKING**
the breakup of concrete due to freeze-thaw expansive pressures within certain aggregates. Also called "D" cracking.

**EDGE CRACKING**
fracture and materials loss in pavements without paved shoulders which occurs along the pavement perimeter. Caused by soil movement beneath the pavement

**EXTRUSION**
to be forced out (i.e., joint sealant from joint)

**FATIGUE CRACKING**
a series of small, jagged, interconnecting cracks caused by failure of the asphalt concrete surface under repeated traffic loading (also called alligator cracking)

**FAULT**
difference in elevation between opposing sides of a joint or crack

**FREE EDGE**
pavement border that is able to move freely

**HAIRLINE CRACK**
a fracture that is very narrow in width, less than 3 mm (0.125 in.)

**JOINT SEAL DAMAGE**
any distress associated with the joint sealant, or lack of joint sealant

**LANE LINE**
boundary between travel lanes, usually a painted stripe

**LANE-TO-SHOULDER DROPOFF**
the difference in elevation between the traffic lane and shoulder

**LANE-TO-SHOULDER SEPARATION**
widening of the joint between the traffic lane and the shoulder

**LEAVE SLAB**
section of pavement just past joint, crack, or other significant roadway feature relative to the direction of traffic

**LONGITUDINAL**
parallel to the centerline of the pavement

**MAP CRACKING**
a series of interconnected hairline cracks in portland cement concrete pavements that extend only into the upper surface of the concrete. Includes cracking typically associated with alkali-silica reactivity (ASR)

**PATCH**
an area where the pavement has been removed and replaced with a new material

**PATCH DETERIORATION**
distress occurring within a previously repaired area

**POLISHED AGGREGATE**
surface mortar and texturing worn away to expose coarse aggregate in the concrete

**POPOUTS**
small pieces of pavement broken loose from the surface

**POTHOLE**
a bowl-shaped depression in the pavement surface

**PUMPING**
the ejection of water and fine materials through cracks in the pavement under moving loads

**PUNCHOUT**
a localized area of a continuously reinforced concrete pavement bounded by two transverse cracks and a longitudinal crack. Aggregate interlock decreases over time and eventually is lost, which leads to steel rupture, and allows the pieces to be punched down into the subbase and subgrade.

**RAVELING**
the wearing away of the pavement surface caused by the dislodging of aggregate particles

**REFLECTION CRACKING**
the fracture of asphalt concrete above joints in the underlying jointed concrete pavement layer(s)

**RUTTING**
longitudinal surface depressions in the wheelpaths

**SCALING**
the deterioration of the upper 3 to 12 mm (0.125 to 0.5 in) of the concrete surface, resulting in the loss of surface mortar

**SHOVING**
permanent, longitudinal displacement of a localized area of the pavement surface caused by traffic pushing against the pavement

**SPALLING**
cracking, breaking, chipping, or fraying of the concrete slab surface within 0.6 m (2 ft) of a joint or crack

**TRANSVERSE**
perpendicular to the pavement centerline

**WATER BLEEDING**
seepage of water from joints or cracks

**WEATHERING**
the wearing away of the pavement surface caused by the loss of asphalt binder

## TABLE OF CONTENTS

**A**

**MANUAL
FOR
DISTRESS
SURVEYS**

## INTRODUCTION

This appendix provides instructions, data sheets, and distress maps for use in visual surveys for the collection of distress information for pavements with asphalt concrete (ACP), jointed concrete (JCP), and continuously reinforced concrete (CRCP) surfaces. The visual distress survey procedures are intended to be used as a back-up at times when it is not possible to schedule the distress contractor vehicle. If the distress contractor has surveyed the test section within three months prior to maintenance and/or rehabilitation work, it will not be necessary to perform the visual distress survey. The visual distress survey will also be performed in remote areas not accessible to the distress contractor (e.g., Hawaii, Puerto Rico).

The *Distress Identification Manual for the Long-Term Pavement Performance Project* shall be used as the standard guide for interpretation, identification, and rating of observed distresses.

During the visual distress survey, safety is the first consideration, as with all field data collection activities. All raters must adhere to the practices and authority of the state.

## EQUIPMENT FOR DISTRESS SURVEYS

The following equipment is necessary for performing field distress surveys of any pavement surface type.

- This field manual
- Extra blank data sheets and maps
- Pencils
- *Distress Identification Manual*
- Clipboard
- Two tape measures, one at least 30 m (100 ft) long and a scale or ruler graduated in millimeters (hundredths of an inch to the nearest 0.05 in.)
- Calculator
- Hard hat and safety vest
- Faultmeter and manual
- 35 mm camera, film
- Video camera, tapes

Transverse profile measurements are required on AC pavements. Additional equipment needed for AC surfaced pavements ONLY, consists of:

- Dipstick and manual.

## INSTRUCTIONS FOR COMPLETING DISTRESS MAPS

The distress maps are used to show the exact location of each distress type existing on the test section. The distress types and severity levels should be identified by using the *Distress Identification Manual.* A total of five sheets are used to map; each sheet contains two 15.25-m (50-ft) maps which represent 30.5 m (100 ft) of the test section (with the exception of SPS-6 sections 2 and 5 which are 300 m [1,000 ft]).

Each test section must be laid out consistently each time a survey is conducted. Sections begin and end at the stations marked on the pavement. Lateral extent of the section, for survey purposes, will vary depending on the existence of longitudinal joints and cracks and the relative position of the lane markings. Figures 1 and 2 illustrate the rules to follow when determining the lateral extent of the section for a distress survey.

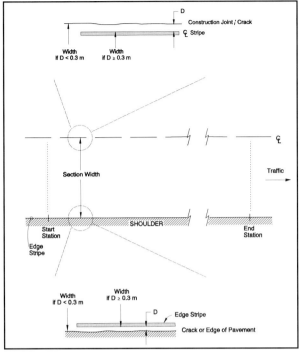

**FIGURE 1**
Test Section Limits for Surveys — Asphalt Surface

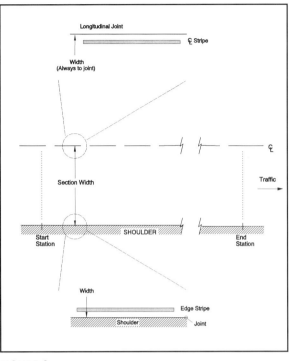

**FIGURE 2**
Test Section Limits for Surveys — Concrete Surface

To map the test section, the tape measure may be placed on the shoulder adjacent to the test section from Station 0 +00 to Station 1+00 (0 to 30.5 m). (An alternative to the tape would be a measuring wheel.) It may be necessary to secure the tape into the pavement with adhesive tape or a heavy object. Once the tape is in place, the distresses can be mapped with the longitudinal placement of the distresses read from the tape. The transverse placement and extent of the distresses can be recorded using the additional tape measure. Once the first 30.5-m (100-ft) subsection is mapped, the tape measure should be moved to map the second 30.5-m (100-ft) subsection. The process is repeated throughout the test section.

The distresses are drawn on the map at the scaled location using the symbols appropriate to the pavement type. In general, the distress is drawn and is labeled using the distress type number and the severity level (L, M, or H) if applicable. For example, a high severity longitudinal crack in the wheel path of an asphalt concrete pavement would be labeled "4aH". An additional symbol is added beside the distress type and severity symbol in cases where the crack or joint is well sealed. Figures specifying the symbols to be used for each pavement type are presented in the following chapters. In addition, example maps are provided to illustrate properly completed maps.

Any observed distresses that are not described in the *Distress Identification Manual* should be photographed and videotaped. The location and extent of the distress should be shown and labeled on the map. Crack sealant and joint sealant condition is to be mapped only for those distresses indicated in Figures 4, 5, and 8. The specific distress types that are not to be included on the maps are to be recorded as follows:

**Appendix A**

**Asphalt Concrete-Surfaced Pavement**

If raveling, polished aggregate, or bleeding occur in large areas over the test section, do not map the total extent. Instead, note the location, extent, and severity level, if applicable, in the space for comments underneath the appropriate map(s). These distresses should be mapped only if they occur in localized areas.

**Jointed Concrete Pavement and Continuously Reinforced Concrete Pavement**

If numerous popouts, map cracking/scaling, or polished aggregate occur in large areas over the test section, do not map the total extent. Instead, note the location, extent, and severity level if applicable in the space for comments underneath the appropriate map(s). These distresses should be mapped only if they occur in localized areas.

## DATA ELEMENTS COMMON TO ALL SURVEY SHEETS

In the common data section appearing in the upper right-hand corner of each of the distress survey data sheets, the four-digit State ID is entered, along with the six-digit SHRP ID (two-digit State code plus four-digit SHRP Section ID). The date the survey was conducted, the initials of up to three raters, before and after pavement surface temperature readings, and the code indicating whether photographs and/or video tape were obtained at the time of the survey are entered in the appropriate spaces.

## INSTRUCTIONS FOR COMPLETING ACP DISTRESS SURVEY SHEETS

Location of the vehicle wheel paths is critical for distinguishing between types of longitudinal cracking in ACP. Figure 3 illustrates the procedure for establishing the location and extent of the wheel paths. Both wheel paths MUST be drawn and identified on the distress maps. The distresses observed are recorded to scale on map sheets. The individual distresses and severity levels depicted on the map are carefully scaled and summed to arrive at the appropriate quantities (e.g., square meters or number of occurrences) and are then recorded on Sheets 1-3. It is important to carefully evaluate the distress map for certain distress types which

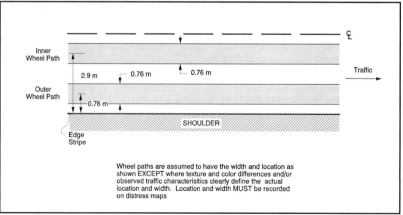

**FIGURE 3**

**Locating Wheel Paths in Asphalt Concrete-Surfaced Pavements**

have multiple methods of measurement due to orientation or location within the section. Reflection cracking at joints, either transverse or longitudinal, and longitudinal cracking, either in the wheel path or elsewhere, are examples of these. Except where indicated otherwise, entries are made for all distress data elements. If a particular type of distress does not exist on the pavement enter "0" as a positive indication that the distress was not overlooked in summarizing the map sheets. All data sheets are to be completed in the field prior to departing the site. Symbols to be used for mapping ACP sections are contained in Figure 4 and an example mapped section is shown in Figure 5.

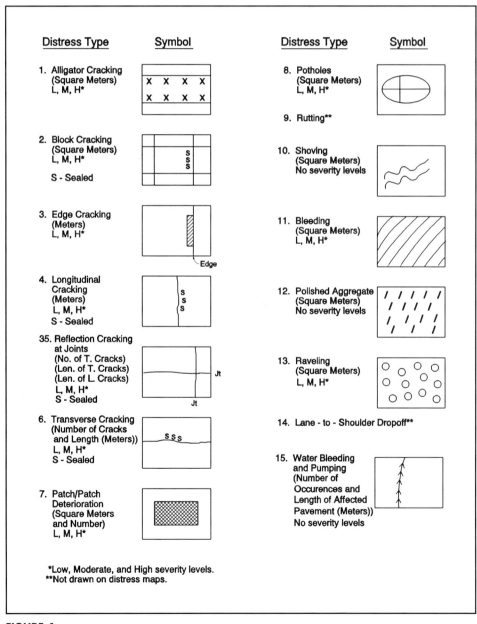

*Low, Moderate, and High severity levels.
**Not drawn on distress maps.

**FIGURE 4**
**Distress Map Symbols for Asphalt Concrete-Surfaced Pavements**

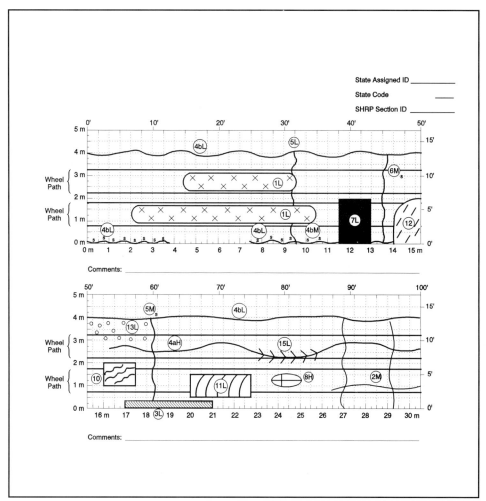

**FIGURE 5**
**Example Map of First 30.5 m (100 ft) of Asphalt Concrete Pavement Section**

### Description of Data Sheet 1

This data sheet provides space for recording measured values for the distress types identified in the left column. The units of measurement for each of the distress types are also identified in the left column. The extent of the measured distress for each particular level of severity is entered in the severity level columns identified as low, moderate, or high. Enter "0" for any distress types and/or severity levels not found.

### Description of Data Sheet 2

This sheet is a continuation of the distress survey data recorded on Sheet 1 and is completed as described under Data Sheet 1. In addition, space is provided to list "Other" distress types found on the test section but not listed on Data Sheet 1 or 2.

**DISTRESS**

**SURVEYS**

### Description of Data Sheet 3

This data sheet provides space to record rutting (using a straight edge 1.2 m [4 ft] long) and lane-to-shoulder dropoff. Manual rutting measurements using a straight edge are only taken for visual surveys conducted on SPS-3 experiment sections. Measurements are taken at the beginning of the test section and at 15.25 m (50 ft) intervals. There should be a total of eleven measurements in each wheel path, for a total of 22 measurements on each test section.

Lane-to-shoulder dropoff is measured as the difference in elevation, to the nearest 1 mm (0.4 in.), between the pavement surface and the adjacent shoulder surface. Measurements are taken at the beginning of the test section and at 15.25 m (50 ft) intervals (a total of 11 measurements) at the lane/shoulder interface or joint. All measurements should be obtained using the Faultmeter. A series of 3 measurements should be performed at each location but only the approximate average of those individual readings should be recorded.

Lane-to-shoulder dropoff typically occurs when the outside shoulder settles. Lane-to-shoulder dropoff is assumed to be a positive value, thus it is not necessary to enter the plus (+) sign before each reading. However, heave of the shoulder may occur due to frost action or swelling soil. If heave of the shoulder is present, it should be recorded as a negative (-) value. At each point where there is no lane-to-shoulder dropoff, enter zero.

## INSTRUCTIONS FOR COMPLETING JCP DATA SHEETS

The distresses observed are recorded to scale on map sheets. This information is reduced by the rater in the field to summarize the results which are then recorded on Sheets 4-7. Except where indicated otherwise, entries are made for all distress data elements. If a particular type of distress does not exist on the pavement enter "0" as a positive indication that the distress was not overlooked in summarizing the map sheets. Symbols to be used for mapping distresses in JCP sections are shown in Figure 6 and an example mapped section is presented in Figure 7.

### Description of Data Sheet 4

This data sheet provides space for recording measured values for the distress types identified in the left column. The units of measurement for each of the distress types are also identified in the left column. The extent of the measured distress for each particular level of severity is entered in the severity level columns identified as low, moderate, or high. Enter zero for any distress types and/or severity levels not found. The distress types and severity levels should be identified by using the *Distress Identification Manual*.

### Description of Data Sheet 5

This sheet is a continuation of the distress survey data recorded on Sheet 4 and is completed as described under Data Sheet 4. In addition, space is provided to list "Other" distress types found on the test section but not listed on Data Sheet 4 or 5.

### Description of Data Sheet 6

This data sheet provides space to record information for each transverse joint and transverse crack encountered in the section. Distance from the beginning of the

**Appendix A**

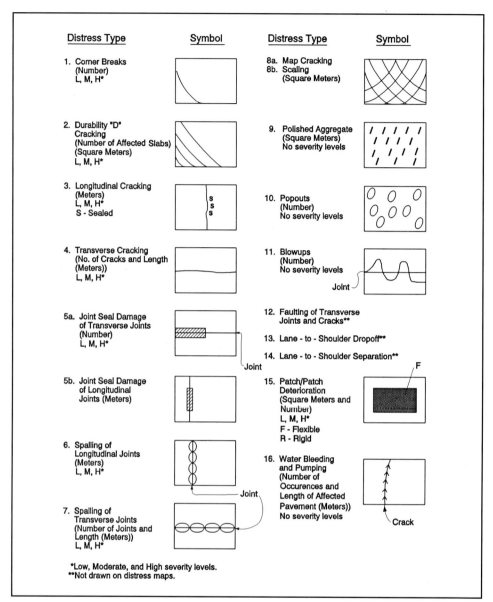

**FIGURE 6**
**Distress Map Symbols for Jointed Concrete Pavements**

section, type of feature (joint or crack), if the feature is a crack then length is recorded, crack sealant condition, and length of joint spalling (crack spalling is not separately recorded). Faulting measurements are made at two transverse locations, 0.3 m (1 ft) and 0.75 m (2.5 ft) from the outside edge of the pavement. At each location three measurements are made but only the approximate average of the readings is recorded to the nearest millimeter. Additional sheets may be required to summarize all joints and transverse cracks in a section so the page number should be recorded in the space provided.

Although no field is provided in the space to the left of the entry for measured faulting, there is room for a negative sign when negative faulting is observed. If the "approach" slab is higher than the "departure" slab, a positive sign is assumed but no entry is required. If the approach slab is lower, a negative sign is entered.

**DISTRESS
SURVEYS**

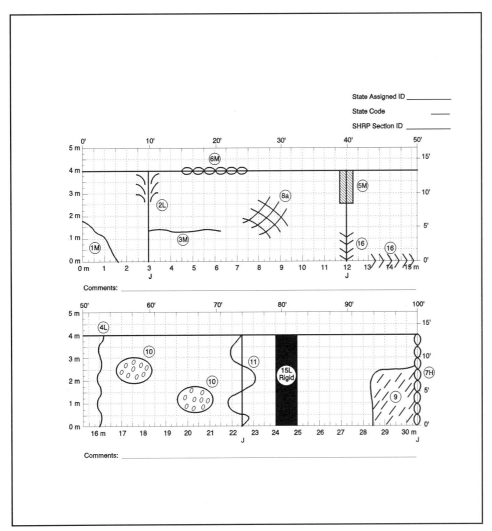

State Assigned ID _____
State Code ___
SHRP Section ID _____

Comments: _____

Comments: _____

**FIGURE 7**
**Example Map of First 30.5 m (100 ft) of a Jointed Concrete Pavement Section**

**Description of Data Sheet 7**

This sheet is used to record lane-to-shoulder dropoff and lane-to-shoulder separa-tion. Lane-to-shoulder dropoff is measured as the difference in elevation, to the nearest 1 mm (0.04 in.), between the pavement surface and the adjacent shoulder surface. Measurements are taken at the beginning of the test section and at 15.25 m (50 ft) intervals (a total of 11 measurements) at the lane/shoulder inter-face or joint. Lane-to-shoulder dropoff typically occurs when the outside shoulder settles. However, heave of the shoulder may occur due to frost action or swelling soil. If heave of the shoulder is present, it should be recorded as a negative (-) value. At each point where there is no lane-to-shoulder dropoff, enter zero.

Lane-to-shoulder separation is measured as the width of the joint (to the nearest 1 mm [0.04 in.]) between the outside lane and the adjacent shoulder surface. Measurements are taken at the beginning of the test section and at 15.25 m (50 ft) intervals (a total of 11 measurements). At each point where there is no lane-to-shoulder separation, enter zero.

**Appendix A**

99

# INSTRUCTIONS FOR COMPLETING CRCP DATA SHEETS

The results of distress surveys on CRCP surfaces are recorded on Sheets 8-10. Except where indicated otherwise, entries are made for all distress data elements. If a particular type of distress does not exist on the pavement enter "0" as a positive indication that the distress was not overlooked in summarizing the map sheets. All data sheets are to be completed in the field prior to departing the site. Symbols to be used for mapping CRCP distresses are contained in Figure 8 and an example mapped section is presented in Figure 9.

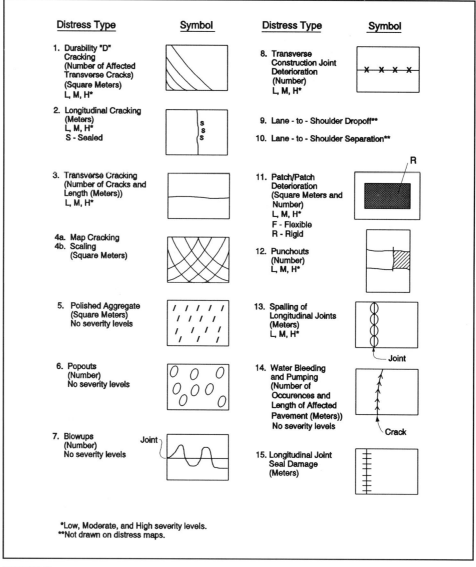

**FIGURE 8**

**Distress Map Symbols for Continuously Reinforced Concrete Pavements**

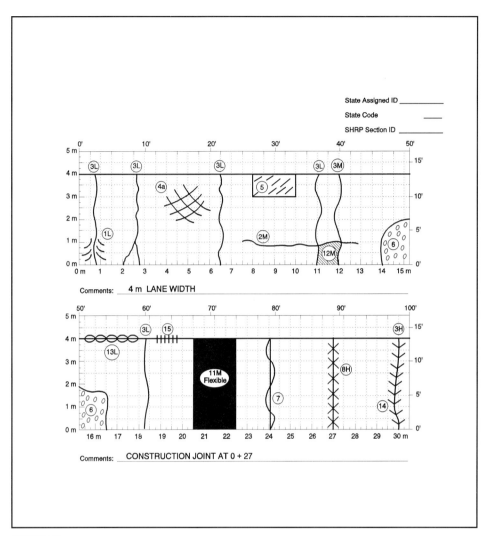

**FIGURE 9**
**Example Map of First 30.5 m (100 ft) of a Continuously Reinforced Concrete
Pavement Section**

## Description of Data Sheet 8

This data sheet provides space for recording measured values for the distress types
identified in the left column. The units of measurement for each of the distress
types are also identified in the left column. The extent of the measured distress
for each particular level of severity is entered in the severity level columns identified as low, moderate, or high, except as indicated on the form. Enter zero for
any distress types and/or severity levels not found. The distress types and severity
levels should be identified by using the *Distress Identification Manual.*

## Description of Data Sheet 9

This sheet is a continuation of the distress survey data recorded on Sheet 8 and is
completed as described under Data Sheet 9. In addition, space is provided to list
"Other" distress types found on the test section but not listed on Data Sheet 8 or 9.

**Appendix A**

101

**Description of Data Sheet 10**

This data sheet provides space to record lane-to-shoulder dropoff and lane-to-shoulder separation. Measurements are taken at the beginning of the test section and at 15.25 m (50 ft) intervals (a total of 11 measurements for each distress) at the lane/shoulder interface or joint.

Lane-to-shoulder dropoff is measured as the difference in elevation (to the nearest 1 mm [0.04 in.]) between the pavement surface and the adjacent shoulder surface. Lane-to-shoulder dropoff typically occurs when the outside shoulder settles. However, heave of the shoulder may occur due to frost action or swelling soil. If heave of the shoulder is present, it should be recorded as a negative (-) value.

Lane-to-shoulder separation is measured as the width of the joint (to the nearest 1 mm [0.04 in.]) between the outside lane and the adjacent shoulder surface.

At each point where there is no lane-to-shoulder dropoff or lane-to-shoulder separation, enter zero.

This part of the appendix shows completed maps and survey forms for a jointed concrete pavement 60 m (197 ft) in length. The rater used the definitions from the *Distress Identification Manual* and the symbols from this appendix when mapping the section. The rater then quantified each distress (and severity levels for the appropriate distresses) on the map. The rater has used the margins of the map sheets to tally the quantities of each distress type. This method makes it easier to total the various distress types, and reduces errors.

The rater then used the tallies to add up the distress quantities, and wrote in the numbers in the appropriate blanks on the survey forms. These forms provide a summary of the distresses found in the JCP section.

**Example
Survey
Maps and
Completed
Sheets**

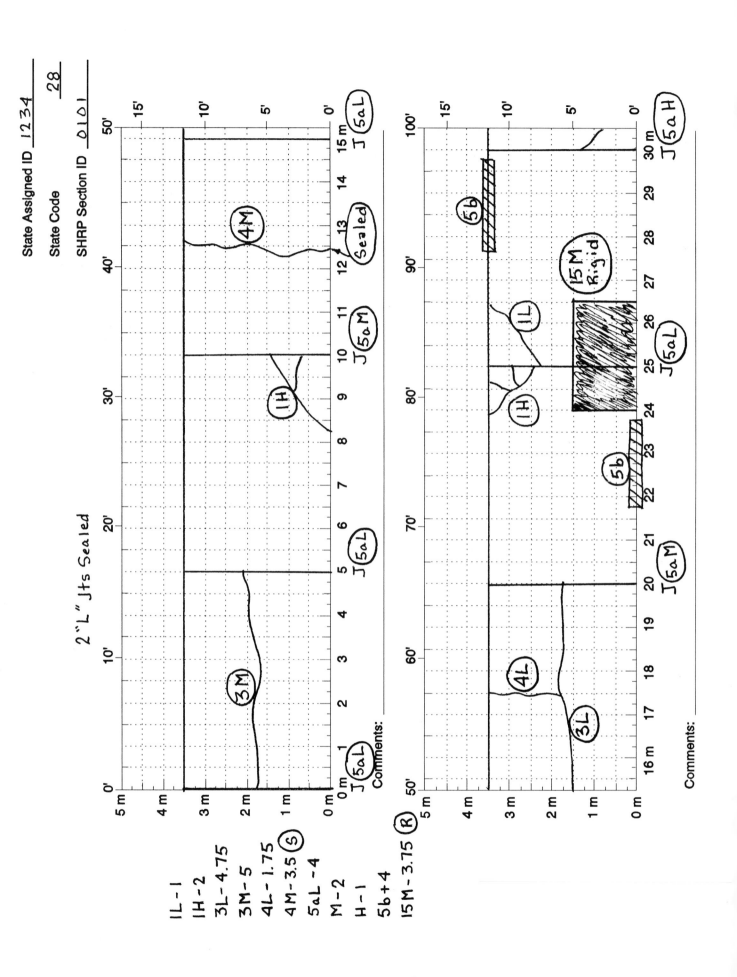

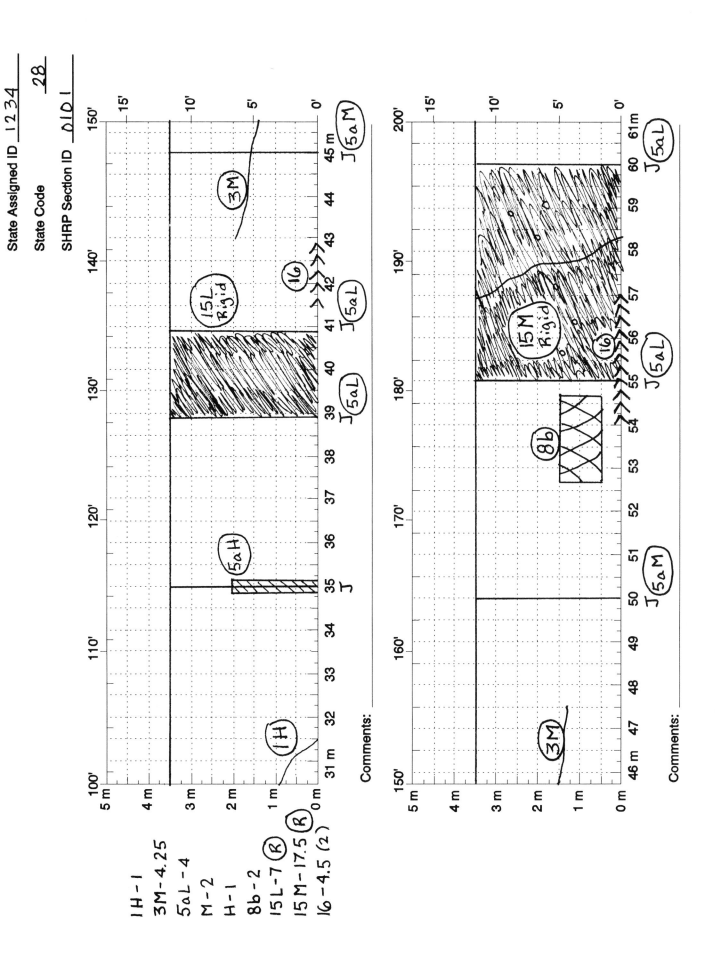

State Assigned ID 1234

State Code 28

SHRP Section ID 0101

Comments:

1H - 1
3M - 4.25
5aL - 4
M - 2
H - 1
8b - 2
15L - 7 (R)
15M - 17.5 (R)
16 - 4.5 (2) (R)

SHEET 4

DISTRESS SURVEY

LTPP PROGRAM

STATE ASSIGNED ID _1_ _2_ _3_ _4_

STATE CODE _2_ _8_

SHRP SECTION ID _0_ _1_ _0_ _1_

## DISTRESS SURVEY FOR PAVEMENTS WITH JOINTED PORTLAND CEMENT CONCRETE SURFACES

DATE OF DISTRESS SURVEY (MONTH/DAY/YEAR)   _0_ _6_/_1_ _2_/_9_ _2_

SURVEYORS: _J S R_, _E J F_, _ _ _

PAVEMENT SURFACE TEMP - BEFORE _ _ _1_ _8_ °C; AFTER _ _ _1_ _9_ °C

PHOTOS, VIDEO, OR BOTH WITH SURVEY (P, V, B) _P_

|  |  | SEVERITY LEVEL | | |
|---|---|---|---|---|
| DISTRESS TYPE | | LOW | MODERATE | HIGH |

**CRACKING**

| | | LOW | MODERATE | HIGH |
|---|---|---|---|---|
| 1. | CORNER BREAKS (Number) | _ _ 1 | _ _ 0 | _ _ 3 |
| 2. | DURABILITY "D" CRACKING (Number of Affected Slabs) | _ _ 0 | _ _ 0 | _ _ 0 |
| | AREA AFFECTED (Square Meters) | _ _ 0.0 | _ _ 0.0 | _ _ 0.0 |
| 3. | LONGITUDINAL CRACKING (Meters) | _ _ 4.8 | _ _ 9.2 | _ _ 0.0 |
| | Length Sealed (Meters) | _ _ 0.0 | _ _ 0.0 | _ _ 0.0 |
| 4. | TRANSVERSE CRACKING (Number of Cracks) (Meters) | _ _ 1 / _ _ 1.8 | _ _ 1 / _ _ 3.5 | _ _ 0 / _ _ 0.0 |
| | Length Sealed (Meters) | _ _ 0.0 | _ _ 3.5 | _ _ 0.0 |

**JOINT DEFICIENCIES**

| | | LOW | MODERATE | HIGH |
|---|---|---|---|---|
| 5a. | TRANSVERSE JOINT SEAL DAMAGE  Sealed? (Y, N)  If "Y" Number of Joints | _ 8 | _ 4 | Y / _ 2 |
| 5b. | LONGITUDINAL JOINT SEAL DAMAGE | | | |
| | Number of Longitudinal Joints that have been sealed (0, 1, or 2) | | | _ 2 |
| | Length of Damaged Sealant (Meters) | | | _ _ 4.0 |
| 6. | SPALLING OF LONGITUDINAL JOINTS (Meters) | _ _ 0.0 | _ _ 0.0 | _ _ 0.0 |
| 7. | SPALLING OF TRANSVERSE JOINTS  Number of Affected Joints  Length Spalled (Meters) | _ 0 / _ _ 0.0 | _ 0 / _ _ 0.0 | _ 0 / _ _ 0.0 |

Revised May 29, 1992

SHEET 5

DISTRESS SURVEY

LTPP PROGRAM

STATE ASSIGNED ID  1 2 3 4

STATE CODE  2 8

SHRP SECTION ID  0 1 0 1

DATE OF DISTRESS SURVEY (MONTH/DAY/YEAR)  0 6/1 2/9 2

SURVEYORS: J S R, E J F

## DISTRESS SURVEY FOR PAVEMENTS WITH JOINTED PORTLAND CEMENT CONCRETE SURFACES (CONTINUED)

| DISTRESS TYPE | SEVERITY LEVEL | | |
|---|---|---|---|
| | LOW | MODERATE | HIGH |

**SURFACE DEFORMATION**

8a. MAP CRACKING (Number) — — 0
(Square Meters) — — 0.0

8b. SCALING (Number) — — 1
(Square Meters) — — 2.0

9. POLISHED AGGREGATE
(Square Meters) — — 0.0

10. POPOUTS (Number per Square Meter) — — — 0

**MISCELLANEOUS DISTRESSES**

11. BLOWUPS (Number) — — 0

12. FAULTING OF TRANSVERSE JOINTS AND CRACKS - REFER TO SHEET 6

13. LANE-TO-SHOULDER DROPOFF - REFER TO SHEET 7

14. LANE-TO-SHOULDER SEPARATION - REFER TO SHEET 7

15. PATCH/PATCH DETERIORATION
Flexible
(Number) — — 0 | — — 0 | — — 0
(Square Meters) — — 0.0 | — — 0.0 | — — 0.0
Rigid
(Number) — — 1 | — — 2 | — — 0
(Square Meters) — — 7.0 | — 2 1.3 | — — 0.0

16. WATER BLEEDING AND PUMPING
(Number of Occurrences) — — 2
Length Affected
(Meters) — — 4.5

17. OTHER (Describe) _____

Revised April 23, 1993

SHEET 6

STATE ASSIGNED ID  _1_ _2_ _3_ _4_

DISTRESS SURVEY

STATE CODE  _2_ _8_

LTPP PROGRAM

SHRP SECTION ID  _0_ _1_ _0_ _1_

DATE OF DISTRESS SURVEY (MONTH/DAY/YEAR) _0_ _6_ / _1_ _2_ / _9_ _2_

SURVEYORS: _J_ _S_ _R_, _E_ _J_ _F_

## DISTRESS SURVEY FOR PAVEMENTS WITH JOINTED PORTLAND CEMENT CONCRETE SURFACES (CONTINUED)

12. FAULTING OF TRANSVERSE JOINTS AND CRACKS

Page _1_ of _1_

| Point[1] Distance (Meters) | Joint or Crack (J/C) | Crack Length (Meters) | Well Sealed (Y/N) | Length of Joint Spalling, m L | M | H | Faulting[2], mm 0.3m | 0.75m |
|---|---|---|---|---|---|---|---|---|
| _ _ 0.1 | J | _._ | _ | 0._ | 0._ | 0._ | _ _ 0 | _ _ 0 |
| _ 5.0 | J | _._ | _ | 0._ | 0._ | 0._ | _ _ 3 | _ _ 4 |
| _ 10.0 | J | _._ | _ | 0._ | 0._ | 0._ | _ _ 2 | _ _ 3 |
| _ 12.3 | C | 3.5 | Y | _._ | _._ | _._ | _ _ 2 | _ _ 1 |
| _ 15.0 | J | _._ | _ | 0._ | 0._ | 0._ | _ _ 1 | _ _ 2 |
| _ 20.0 | J | _._ | _ | 0._ | 0._ | 0._ | _ _ 4 | _ _ 5 |
| _ 25.0 | J | _._ | _ | 0._ | 0._ | 0._ | _ _ 3 | _ _ 2 |
| _ 30.0 | J | _._ | _ | 0._ | 0._ | 0._ | _ _ L | _ _ 0 |
| _ 35.0 | J | _._ | _ | 0._ | 0._ | 0._ | _ _ 1 | _ _ 0 |
| _ 38.8 | J | _._ | _ | 0._ | 0._ | 0._ | _ _ 5 | _ - 4 |
| _ 40.8 | J | _._ | _ | 0._ | 0._ | 0._ | _ _ 3 | _ - 4 |
| _ 45.0 | J | _._ | _ | 0._ | 0._ | 0._ | _ _ 2 | _ _ 3 |
| _ 50.0 | J | _._ | _ | 0._ | 0._ | 0._ | _ _ 1 | _ _ 0 |
| _ 55.0 | J | _._ | _ | 0._ | 0._ | 0._ | _ _ 1 | _ _ 0 |
| _ 60.0 | J | _._ | _ | 0._ | 0._ | 0._ | _ _ 0 | _ _ 1 |
| _ _ _._ | _ | _._ | _ | _._ | _._ | _._ | _ _ _ | _ _ _ |
| _ _ _._ | _ | _._ | _ | _._ | _._ | _._ | _ _ _ | _ _ _ |
| _ _ _._ | _ | _._ | _ | _._ | _._ | _._ | _ _ _ | _ _ _ |
| _ _ _._ | _ | _._ | _ | _._ | _._ | _._ | _ _ _ | _ _ _ |
| _ _ _._ | _ | _._ | _ | _._ | _._ | _._ | _ _ _ | _ _ _ |
| _ _ _._ | _ | _._ | _ | _._ | _._ | _._ | _ _ _ | _ _ _ |
| _ _ _._ | _ | _._ | _ | _._ | _._ | _._ | _ _ _ | _ _ _ |
| _ _ _._ | _ | _._ | _ | _._ | _._ | _._ | _ _ _ | _ _ _ |
| _ _ _._ | _ | _._ | _ | _._ | _._ | _._ | _ _ _ | _ _ _ |
| _ _ _._ | _ | _._ | _ | _._ | _._ | _._ | _ _ _ | _ _ _ |
| _ _ _._ | _ | _._ | _ | _._ | _._ | _._ | _ _ _ | _ _ _ |

Note 1.  Point Distance is from the start of the test section to the measurement location.

Note 2.  If the "approach" slab is higher than the "departure" slab, faulting is recorded as positive (+ or 0); if the "approach" slab is lower, record faulting as negative (-) and the minus sign must be used.

STATE ASSIGNED ID  1 2 3 4

SHEET 7

DISTRESS SURVEY                    STATE CODE          2 8

LTPP PROGRAM                       SHRP SECTION ID   0 1 0 1

DATE OF DISTRESS SURVEY (MONTH/DAY/YEAR) 0 6/1 2/9 2
SURVEYORS: J S R, E J E

## DISTRESS SURVEY FOR PAVEMENTS WITH JOINTED
## PORTLAND CEMENT CONCRETE SURFACES
## (CONTINUED)

13. LANE-TO-SHOULDER DROPOFF

14. LANE-TO-SHOULDER SEPARATION

| Point No. | Point[1] Distance (meters) | Lane-to-shoulder[2] Dropoff (mm) | Lane-to-shoulder Separation (mm) | Well Sealed (Y/N) |
|---|---|---|---|---|
| 1. | 0. | 4. | 8. | Y |
| 2. | 15.25 | 8. | 6. | Y |
| 3. | 30.5 | 0. | 10. | Y |
| 4. | 45.75 | 6. | 8. | Y |
| 5. | 61. | . | . | |
| 6. | 76.25 | . | . | |
| 7. | 91.5 | . | . | |
| 8. | 106.75 | . | . | |
| 9. | 122. | . | . | |
| 10. | 137.25 | . | . | |
| 11. | 152.5 | . | . | |

Not Mapped

Note 1.    Point Distance is from the start of the test section to the measurement location. The values shown are SI equivalents of the 50 ft spacing used in previous surveys.

Note 2.    If heave of the shoulder occurs (upward movement), record as a negative (-) value. Do not record (+) signs, positive values are assumed.

These map forms and data sheets may be photocopied from this book for use in the field. Note that each type of pavement has its own data sheets.

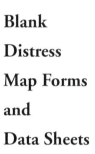

| | | |
|---|---|---|
| ACP | sheets **1, 2, 3** | pages **117, 118, 119** |
| JCP | sheets **4, 5, 6, 7** | pages **120, 121, 122, 123** |
| CRCP | sheets **8, 9, 10** | pages **124, 125, 126** |

**Blank**

**Distress**

**Map Forms**

**and**

**Data Sheets**

State Assigned ID _____

State Code _____

SHRP Section ID _____

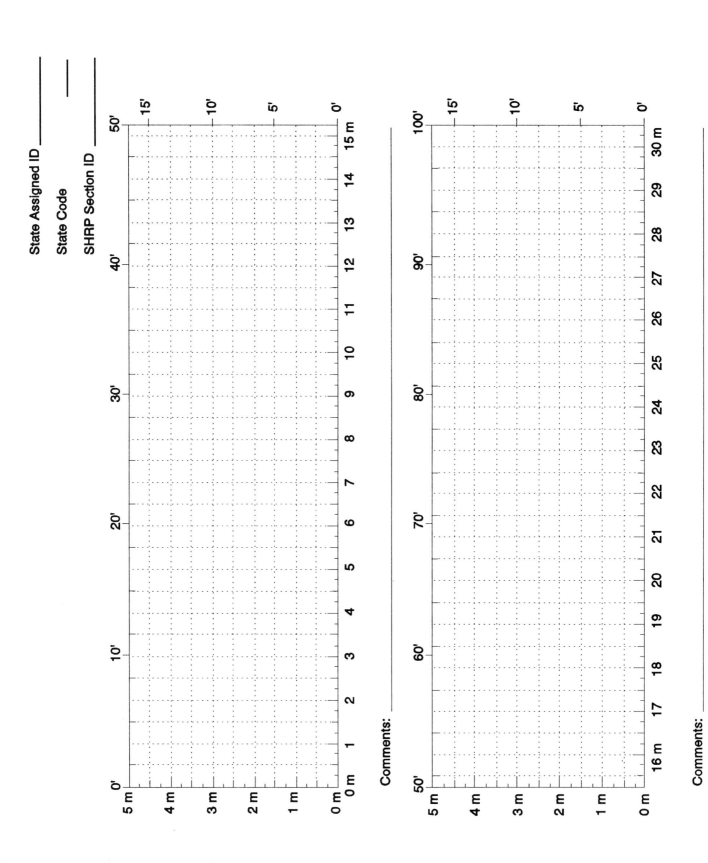

Comments: _____

Comments: _____

State Assigned ID _____

State Code _____

SHRP Section ID _____

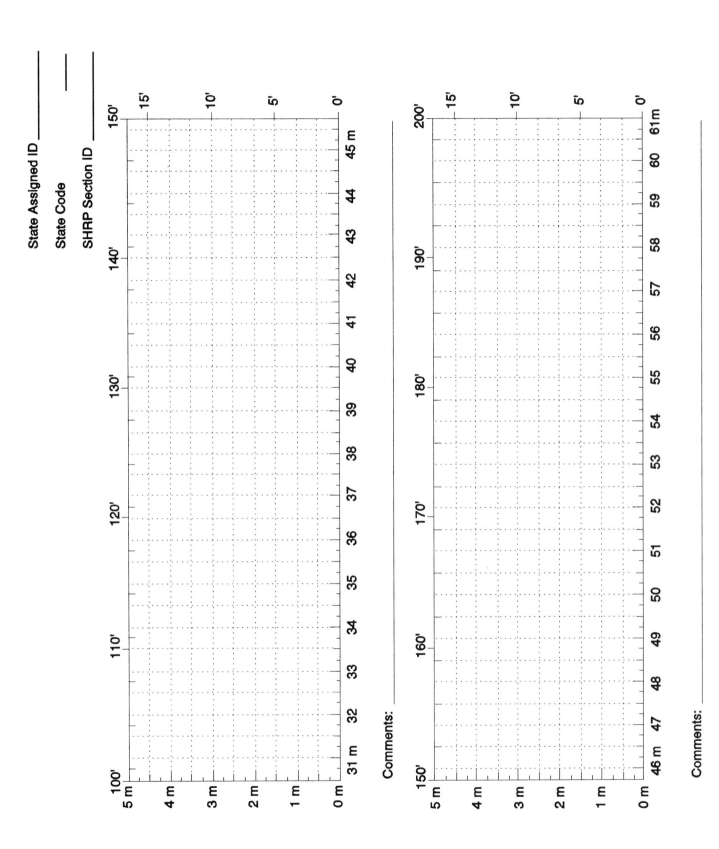

Comments: _____

Comments: _____

State Assigned ID _____

State Code _____

SHRP Section ID _____

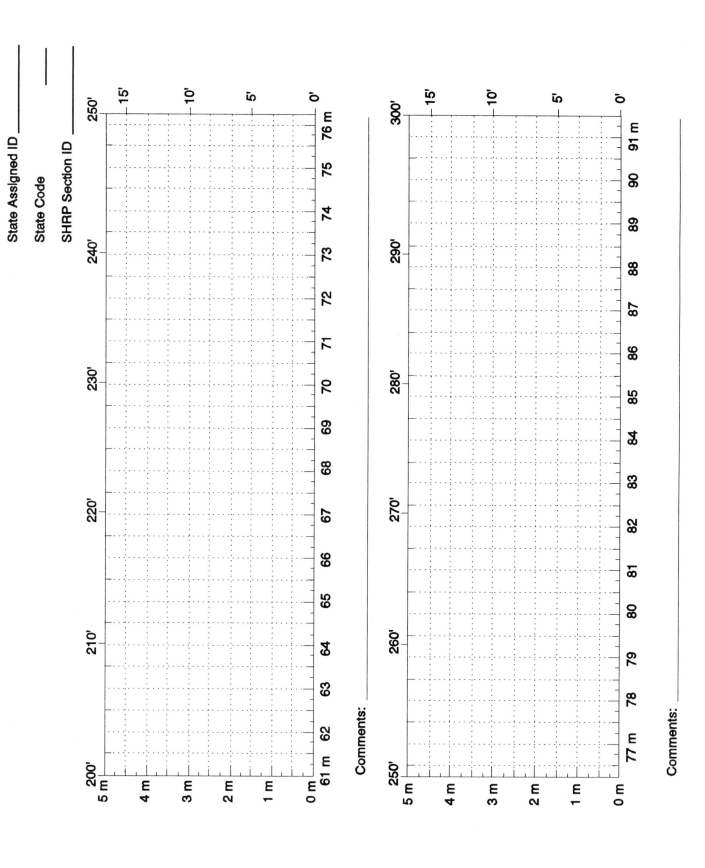

Comments:

Comments:

State Assigned ID _____

State Code ___

SHRP Section ID _____

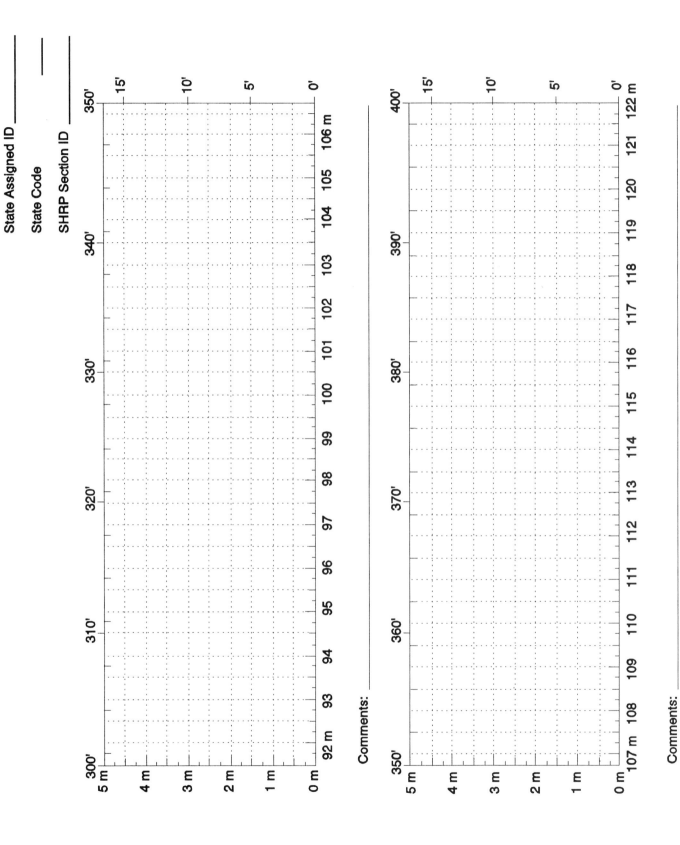

Comments: _____

Comments:

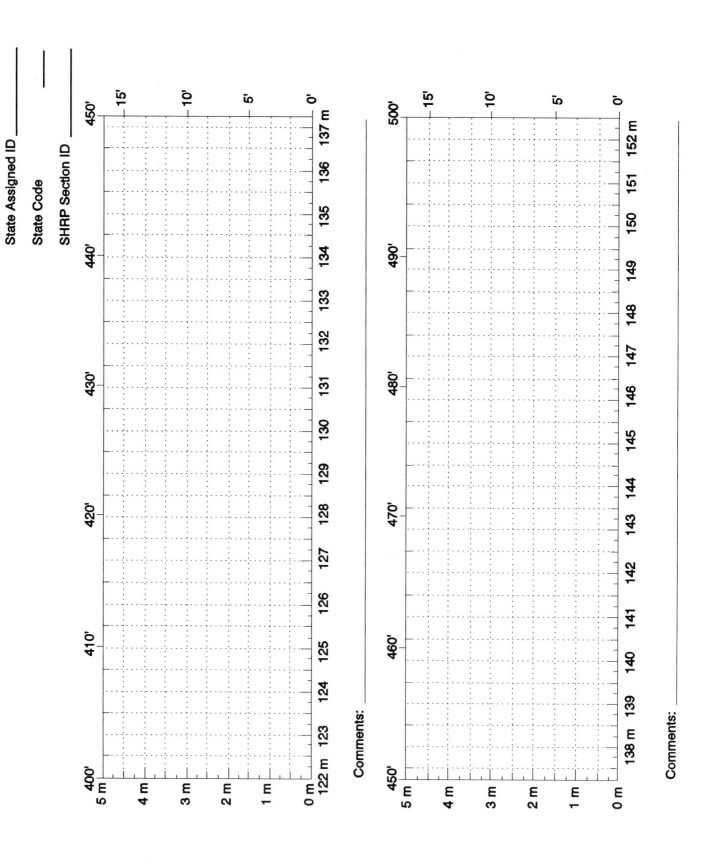

Comments: _____

Comments: _____

STATE ASSIGNED ID  __ __ __ __

SHEET 1

DISTRESS SURVEY

STATE CODE  __ __

LTPP PROGRAM

SHRP SECTION ID  __ __ __ __

<u>DISTRESS SURVEY FOR PAVEMENTS WITH ASPHALT CONCRETE SURFACES</u>

DATE OF DISTRESS SURVEY (MONTH/DAY/YEAR)  __ __/__ __/__ __

SURVEYORS: __ __ __, __ __ __  PHOTOS, VIDEO, OR BOTH WITH SURVEY (P, V, B) __
PAVEMENT SURFACE TEMP - BEFORE __ __ __ __°C; AFTER __ __ __ __°C

| DISTRESS TYPE | SEVERITY LEVEL | | |
| --- | --- | --- | --- |
| | LOW | MODERATE | HIGH |
| **CRACKING** | | | |
| 1. FATIGUE CRACKING (Square Meters) | __ __ __.__ | __ __ __.__ | __ __ __.__ |
| 2. BLOCK CRACKING (Square Meters) | __ __ __.__ | __ __ __.__ | __ __ __.__ |
| 3. EDGE CRACKING (Meters) | __ __ __.__ | __ __ __.__ | __ __ __.__ |
| 4. LONGITUDINAL CRACKING (Meters) | | | |
| 4a. Wheel Path | __ __ __.__ | __ __ __.__ | __ __ __.__ |
| Length Sealed (Meters) | __ __ __.__ | __ __ __.__ | __ __ __.__ |
| 4b. Non-Wheel Path | __ __ __.__ | __ __ __.__ | __ __ __.__ |
| Length Sealed (Meters) | __ __ __.__ | __ __ __.__ | __ __ __.__ |
| 5. REFLECTION CRACKING AT JOINTS Number of Transverse Cracks | __ __ __ | __ __ __ | __ __ __ |
| Transverse Cracking (Meters) | __ __ __.__ | __ __ __.__ | __ __ __.__ |
| Length Sealed (Meters) | __ __ __.__ | __ __ __.__ | __ __ __.__ |
| Longitudinal Cracking (Meters) | __ __ __.__ | __ __ __.__ | __ __ __.__ |
| Length Sealed (Meters) | __ __ __.__ | __ __ __.__ | __ __ __.__ |
| 6. TRANSVERSE CRACKING Number of Cracks | __ __ __ | __ __ __ | __ __ __ |
| Length (Meters) | __ __ __.__ | __ __ __.__ | __ __ __.__ |
| Length Sealed (Meters) | __ __ __.__ | __ __ __.__ | __ __ __.__ |
| **PATCHING AND POTHOLES** | | | |
| 7. PATCH/PATCH DETERIORATION (Number) | __ __ __ | __ __ __ | __ __ __ |
| (Square Meters) | __ __ __.__ | __ __ __.__ | __ __ __.__ |
| 8. Potholes (Number) | __ __ __ | __ __ __ | __ __ __ |
| (Square Meters) | __ __ __.__ | __ __ __.__ | __ __ __.__ |

Revised December 1, 1992

SHEET 2

DISTRESS SURVEY

LTPP PROGRAM

STATE ASSIGNED ID __ __ __ __
_____

STATE CODE __ __

SHRP SECTION ID __ __ __ __
_____

DATE OF DISTRESS SURVEY (MONTH/DAY/YEAR) __ __/__ __/__ __

SURVEYORS: __ __ __, __ __ __

## DISTRESS SURVEY FOR PAVEMENTS WITH ASPHALT CONCRETE SURFACES (CONTINUED)

| DISTRESS TYPE | SEVERITY LEVEL | | |
|---|---|---|---|
| | LOW | MODERATE | HIGH |

**SURFACE DEFORMATION**

9. RUTTING - REFER TO SHEET 3 FOR SPS-3 OR Form S1 from Dipstick Manual

10. SHOVING
(Number)     __ __ __
(Square Meters)     __ __ __.__

**SURFACE DEFECTS**

11. BLEEDING
(Square Meters)    __ __ __.__    __ __ __.__    __ __ __.__

12. POLISHED AGGREGATE
(Square Meters)     __ __ __.__

13. RAVELING
(Square Meters)    __ __ __.__    __ __ __.__    __ __ __.__

**MISCELLANEOUS DISTRESSES**

14. LANE-TO-SHOULDER DROPOFF - REFER TO SHEET 3

15. WATER BLEEDING AND PUMPING
(Number)     __ __ __
Length of Affected Pavement
(Meters)     __ __ __.__

16. OTHER (Describe) _____

_____

_____

_____

SHEET 3

DISTRESS SURVEY

LTPP PROGRAM

STATE ASSIGNED ID  __ __ __ __

STATE CODE  __ __

SHRP SECTION ID  __ __ __ __

DATE OF DISTRESS SURVEY (MONTH/DAY/YEAR)  __ __/__ __/__ __

SURVEYORS:  __ __ __, __ __ __

## DISTRESS SURVEY FOR PAVEMENTS WITH ASPHALT CONCRETE SURFACES (CONTINUED)

9.   RUTTING    (FOR SPS-3 SITE SURVEYS)

| | INNER WHEEL PATH | | | OUTER WHEEL PATH | |
|---|---|---|---|---|---|
| Point No. | Point Distance[1] (Meters) | Rut Depth (mm) | Point No. | Point Distance[1] (Meters) | Rut Depth (mm) |
| 1 | 0. | __ __ __. | 1 | 0. | __ __ __. |
| 2 | 15.25 | __ __ __. | 2 | 15.25 | __ __ __. |
| 3 | 30.5 | __ __ __. | 3 | 30.5 | __ __ __. |
| 4 | 45.75 | __ __ __. | 4 | 45.75 | __ __ __. |
| 5 | 61. | __ __ __. | 5 | 61. | __ __ __. |
| 6 | 76.25 | __ __ __. | 6 | 76.25 | __ __ __. |
| 7 | 91.5 | __ __ __. | 7 | 91.5 | __ __ __. |
| 8 | 106.75 | __ __ __. | 8 | 106.75 | __ __ __. |
| 9 | 122. | __ __ __. | 9 | 122. | __ __ __. |
| 10 | 137.25 | __ __ __. | 10 | 137.25 | __ __ __. |
| 11 | 152.5 | __ __ __. | 11 | 152.5 | __ __ __. |

14.   LANE-TO-SHOULDER DROPOFF

| Point No. | Point Distance[1] Meters | Lane-to-Shoulder Dropoff (mm) |
|---|---|---|
| 1 | 0. | __ __ __. |
| 2 | 15.25 | __ __ __. |
| 3 | 30.5 | __ __ __. |
| 4 | 45.75 | __ __ __. |
| 5 | 61. | __ __ __. |
| 6 | 76.25 | __ __ __. |
| 7 | 91.5 | __ __ __. |
| 8 | 106.75 | __ __ __. |
| 9 | 122. | __ __ __. |
| 10 | 137.25 | __ __ __. |
| 11 | 152.5 | __ __ __. |

Note 1:   "Point Distance" is the distance in meters from the start of the test section to the point where the measurement was made. The values shown are SI equivalents of the 50 ft spacing used in previous surveys.

STATE ASSIGNED ID  __ __ __ __

SHEET 4

STATE CODE  __ __

DISTRESS SURVEY

LTPP PROGRAM

SHRP SECTION ID  __ __ __ __

## DISTRESS SURVEY FOR PAVEMENTS WITH JOINTED PORTLAND CEMENT CONCRETE SURFACES

DATE OF DISTRESS SURVEY (MONTH/DAY/YEAR)  __ __/__ __/__ __

SURVEYORS:  __ __ __ , __ __ __ , __ __ __
PAVEMENT SURFACE TEMP - BEFORE __ __ __ __°C; AFTER __ __ __ __°C
PHOTOS, VIDEO, OR BOTH WITH SURVEY (P, V, B) __

| DISTRESS TYPE | SEVERITY LEVEL | | |
| --- | --- | --- | --- |
| | LOW | MODERATE | HIGH |

### CRACKING

1. CORNER BREAKS (Number) __ __ __ | __ __ __ | __ __ __

2. DURABILITY "D" CRACKING
   (Number of Affected Slabs) __ __ __ | __ __ __ | __ __ __
   AREA AFFECTED
   (Square Meters) __ __ __.__ | __ __ __.__ | __ __ __.__

3. LONGITUDINAL CRACKING
   (Meters) __ __ __.__ | __ __ __.__ | __ __ __.__
   Length Sealed
   (Meters) __ __ __.__ | __ __ __.__ | __ __ __.__

4. TRANSVERSE CRACKING
   (Number of Cracks) __ __ __ | __ __ __ | __ __ __
   (Meters) __ __ __.__ | __ __ __.__ | __ __ __.__

   Length Sealed
   (Meters) __ __ __.__ | __ __ __.__ | __ __ __.__

### JOINT DEFICIENCIES

5a. TRANSVERSE JOINT SEAL DAMAGE
    Sealed?  (Y, N) __
    If "Y" Number of Joints __ __ | __ __ | __ __

5b. LONGITUDINAL JOINT SEAL DAMAGE
    Number of Longitudinal Joints that have been sealed (0, 1, or 2) __
    Length of Damaged Sealant (Meters) __ __ __.__

6. SPALLING OF LONGITUDINAL JOINTS
   (Meters) __ __ __.__ | __ __ __.__ | __ __ __.__

7. SPALLING OF TRANSVERSE JOINTS
   Number of Affected Joints __ __ | __ __ | __ __
   Length Spalled (Meters) __ __ __.__ | __ __ __.__ | __ __ __.__

Revised May 29, 1992

SHEET 5

DISTRESS SURVEY

LTPP PROGRAM

STATE ASSIGNED ID __ __ __ __

STATE CODE __ __

SHRP SECTION ID __ __ __ __

DATE OF DISTRESS SURVEY (MONTH/DAY/YEAR) __ __/__ __/__ __

SURVEYORS: __ __ __ , __ __ __

## DISTRESS SURVEY FOR PAVEMENTS WITH JOINTED PORTLAND CEMENT CONCRETE SURFACES (CONTINUED)

| DISTRESS TYPE | SEVERITY LEVEL | | |
|---|---|---|---|
| | LOW | MODERATE | HIGH |

**SURFACE DEFORMATION**

8a. MAP CRACKING (Number) __ __ __
    (Square Meters) __ __ __.__

8b. SCALING (Number) __ __ __
    (Square Meters) __ __ __.__

9. POLISHED AGGREGATE
   (Square Meters) __ __ __.__

10. POPOUTS (Number per Square Meter) __ __ __ __

**MISCELLANEOUS DISTRESSES**

11. BLOWUPS (Number) __ __ __

12. FAULTING OF TRANSVERSE JOINTS AND CRACKS - REFER TO SHEET 6

13. LANE-TO-SHOULDER DROPOFF - REFER TO SHEET 7

14. LANE-TO-SHOULDER SEPARATION - REFER TO SHEET 7

15. PATCH/PATCH DETERIORATION
    Flexible
        (Number) __ __ __   __ __ __   __ __ __
        (Square Meters) __ __ __.__   __ __ __.__   __ __ __.__
    Rigid
        (Number) __ __ __   __ __ __   __ __ __
        (Square Meters) __ __ __.__   __ __ __.__   __ __ __.__

16. WATER BLEEDING AND PUMPING
    (Number of Occurrences) __ __ __
    Length Affected
    (Meters) __ __ __.__

17. OTHER (Describe) _____

SHEET 6

DISTRESS SURVEY

LTPP PROGRAM

STATE ASSIGNED ID   __ __ __ __
_____

STATE CODE          __ __

SHRP SECTION ID     __ __ __ __
_____

DATE OF DISTRESS SURVEY (MONTH/DAY/YEAR) __ __/__ __/__ __
SURVEYORS:  __ __ __, __ __ __

## DISTRESS SURVEY FOR PAVEMENTS WITH JOINTED PORTLAND CEMENT CONCRETE SURFACES (CONTINUED)

12. FAULTING OF TRANSVERSE JOINTS AND CRACKS                    Page __ of __

| Point[1] Distance (Meters) | Joint or Crack (J/C) | Crack Length (Meters) | Well Sealed (Y/N) | Length of Joint Spalling, m | | | Faulting[2], mm | |
|---|---|---|---|---|---|---|---|---|
| | | | | L | M | H | 0.3m | 0.75m |
| _ _ _·_ | _ | _·_ | _ | _·_ | _·_ | _·_ | _ _ _ | _ _ _ |
| _ _ _·_ | _ | _·_ | _ | _·_ | _·_ | _·_ | _ _ _ | _ _ _ |
| _ _ _·_ | _ | _·_ | _ | _·_ | _·_ | _·_ | _ _ _ | _ _ _ |
| _ _ _·_ | _ | _·_ | _ | _·_ | _·_ | _·_ | _ _ _ | _ _ _ |
| _ _ _·_ | _ | _·_ | _ | _·_ | _·_ | _·_ | _ _ _ | _ _ _ |
| _ _ _·_ | _ | _·_ | _ | _·_ | _·_ | _·_ | _ _ _ | _ _ _ |
| _ _ _·_ | _ | _·_ | _ | _·_ | _·_ | _·_ | _ _ _ | _ _ _ |
| _ _ _·_ | _ | _·_ | _ | _·_ | _·_ | _·_ | _ _ _ | _ _ _ |
| _ _ _·_ | _ | _·_ | _ | _·_ | _·_ | _·_ | _ _ _ | _ _ _ |
| _ _ _·_ | _ | _·_ | _ | _·_ | _·_ | _·_ | _ _ _ | _ _ _ |
| _ _ _·_ | _ | _·_ | _ | _·_ | _·_ | _·_ | _ _ _ | _ _ _ |
| _ _ _·_ | _ | _·_ | _ | _·_ | _·_ | _·_ | _ _ _ | _ _ _ |
| _ _ _·_ | _ | _·_ | _ | _·_ | _·_ | _·_ | _ _ _ | _ _ _ |
| _ _ _·_ | _ | _·_ | _ | _·_ | _·_ | _·_ | _ _ _ | _ _ _ |
| _ _ _·_ | _ | _·_ | _ | _·_ | _·_ | _·_ | _ _ _ | _ _ _ |
| _ _ _·_ | _ | _·_ | _ | _·_ | _·_ | _·_ | _ _ _ | _ _ _ |
| _ _ _·_ | _ | _·_ | _ | _·_ | _·_ | _·_ | _ _ _ | _ _ _ |
| _ _ _·_ | _ | _·_ | _ | _·_ | _·_ | _·_ | _ _ _ | _ _ _ |
| _ _ _·_ | _ | _·_ | _ | _·_ | _·_ | _·_ | _ _ _ | _ _ _ |
| _ _ _·_ | _ | _·_ | _ | _·_ | _·_ | _·_ | _ _ _ | _ _ _ |
| _ _ _·_ | _ | _·_ | _ | _·_ | _·_ | _·_ | _ _ _ | _ _ _ |
| _ _ _·_ | _ | _·_ | _ | _·_ | _·_ | _·_ | _ _ _ | _ _ _ |
| _ _ _·_ | _ | _·_ | _ | _·_ | _·_ | _·_ | _ _ _ | _ _ _ |
| _ _ _·_ | _ | _·_ | _ | _·_ | _·_ | _·_ | _ _ _ | _ _ _ |
| _ _ _·_ | _ | _·_ | _ | _·_ | _·_ | _·_ | _ _ _ | _ _ _ |
| _ _ _·_ | _ | _·_ | _ | _·_ | _·_ | _·_ | _ _ _ | _ _ _ |
| _ _ _·_ | _ | _·_ | _ | _·_ | _·_ | _·_ | _ _ _ | _ _ _ |

Note 1.  Point Distance is from the start of the test section to the measurement location.

Note 2.  If the "approach" slab is higher than the "departure" slab, faulting is recorded as positive (+ or 0); if the "approach" slab is lower, record faulting as negative (-) and the minus sign must be used.

Revised May 29, 1992

SHEET 7

DISTRESS SURVEY

LTPP PROGRAM

STATE ASSIGNED ID _ _ _ _

STATE CODE _ _

SHRP SECTION ID _ _ _ _

DATE OF DISTRESS SURVEY (MONTH/DAY/YEAR) _ _/_ _/_ _
SURVEYORS: _ _ _, _ _ _

**DISTRESS SURVEY FOR PAVEMENTS WITH JOINTED PORTLAND CEMENT CONCRETE SURFACES (CONTINUED)**

13. LANE-TO-SHOULDER DROPOFF

14. LANE-TO-SHOULDER SEPARATION

| Point No. | Point[1] Distance (meters) | Lane-to-shoulder[2] Dropoff (mm) | Lane-to-shoulder Separation (mm) | Well Sealed (Y/N) |
|---|---|---|---|---|
| 1. | 0. | _ _ _. | _ _ _. | _ |
| 2. | 15.25 | _ _ _. | _ _ _. | _ |
| 3. | 30.5 | _ _ _. | _ _ _. | _ |
| 4. | 45.75 | _ _ _. | _ _ _. | _ |
| 5. | 61. | _ _ _. | _ _ _. | _ |
| 6. | 76.25 | _ _ _. | _ _ _. | _ |
| 7. | 91.5 | _ _ _. | _ _ _. | _ |
| 8. | 106.75 | _ _ _. | _ _ _. | _ |
| 9. | 122. | _ _ _. | _ _ _. | _ |
| 10. | 137.25 | _ _ _. | _ _ _. | _ |
| 11. | 152.5 | _ _ _. | _ _ _. | _ |

Note 1.  Point Distance is from the start of the test section to the measurement location. The values shown are SI equivalents of the 50 ft spacing used in previous surveys.

Note 2.  If heave of the shoulder occurs (upward movement), record as a negative (-) value. Do not record (+) signs, positive values are assumed.

SHEET 8

DISTRESS SURVEY

LTPP PROGRAM

STATE ASSIGNED ID  __ __ __ __
_____

STATE CODE  __ __
_____

SHRP SECTION ID  __ __ __ __
_____

<u>DISTRESS SURVEY FOR PAVEMENTS WITH CONTINUOUSLY</u>
<u>REINFORCED PORTLAND CEMENT CONCRETE SURFACES</u>

DATE OF DISTRESS SURVEY (MONTH/DAY/YEAR)          __ __/__ __/__ __

SURVEYORS: __ __ __, __ __ __   PHOTOS, VIDEO, OR BOTH WITH SURVEY (P,V,B) __
PAVEMENT SURFACE TEMP - BEFORE __ __ __ __°C; AFTER __ __ __ __°C

|                          | SEVERITY LEVEL | | |
| --- | --- | --- | --- |
| DISTRESS TYPE | LOW | MODERATE | HIGH |

**CRACKING**

1. DURABILITY "D" CRACKING
(No. of Affected Trans Cracks)   __ __ __        __ __ __        __ __ __
(Square Meters)                  __ __ __.__     __ __ __.__     __ __ __.__

2. LONGITUDINAL CRACKING
(Meters)                         __ __ __.__     __ __ __.__     __ __ __.__
Length Well Sealed
(Meters)                         __ __ __.__     __ __ __.__     __ __ __.__

3. TRANSVERSE CRACKING
(Total Number of Cracks)                                        __ __ __
(Number of Cracks)               __ __ __        __ __ __        __ __ __
(Meters)                         __ __ __.__     __ __ __.__     __ __ __.__

**SURFACE DEFECTS**

4a. MAP CRACKING (Number)                                       __ __ __
(Square Meters)                                                 __ __ __.__

4b. SCALING (Number)                                            __ __ __
(Square Meters)                                                 __ __ __.__

5. POLISHED AGGREGATE
(Square Meters)                                                 __ __ __.__

6. POPOUTS (Number per Square Meter)                            __ __ __

STATE ASSIGNED ID  __ __ __ __

SHEET 9

DISTRESS SURVEY

LTPP PROGRAM

STATE CODE                    __ __

SHRP SECTION ID    __ __ __ __

DATE OF DISTRESS SURVEY (MONTH/DAY/YEAR) __ __/__ __/__ __

SURVEYORS: __ __ __, __ __ __

## DISTRESS SURVEY FOR PAVEMENTS WITH CONTINUOUSLY REINFORCED PORTLAND CEMENT CONCRETE SURFACES (CONTINUED)

| DISTRESS TYPE | SEVERITY LEVEL | | |
|---|---|---|---|
| | LOW | MODERATE | HIGH |

**MISCELLANEOUS DISTRESSES**

7.  BLOWUPS (Number)                                                                         __ __ __

8.  TRANSVERSE CONSTRUCTION JOINT
    DETERIORATION (Number)                        __ __            __ __            __ __

9.  LANE-TO-SHOULDER DROPOFF - REFER TO SHEET 10

10. LANE-TO-SHOULDER SEPARATION - REFER TO SHEET 10

11. PATCH/PATCH DETERIORATION
    Flexible
    (Number)                                      __ __ __        __ __ __        __ __ __
    (Square Meters)                               __ __ __.__     __ __ __.__     __ __ __.__
    Rigid
    (Number)                                      __ __ __        __ __ __        __ __ __
    (Square Meters)                               __ __ __.__     __ __ __.__     __ __ __.__

12. PUNCHOUTS (Number)                            __ __            __ __            __ __

13. SPALLING OF LONGITUDINAL
    JOINT (Meters)                                __ __ __.__     __ __ __.__     __ __ __.__

14. WATER BLEEDING AND PUMPING
    (Number of Occurrences)                                                       __ __ __
    Length Affected
    (Meters)                                                                      __ __ __.__

15. LONGITUDINAL JOINT SEAL DAMAGE
    Number of Longitudinal Joints that have been sealed (0, 1, or 2)              __
    If Sealed Length w/Damaged Sealant (Meters)                                   __ __ __.__

16. OTHER (Describe) _____

    _____

    _____

Revised May 29, 1992

SHEET 10

DISTRESS SURVEY

LTPP PROGRAM

STATE ASSIGNED ID  __ __ __ __

STATE CODE  __ __

SHRP SECTION ID  __ __ __ __

DATE OF DISTRESS SURVEY (MONTH/DAY/YEAR) __ __/__ __/__ __

SURVEYORS: __ __ __, __ __ __

## DISTRESS SURVEY FOR PAVEMENTS WITH CONTINUOUSLY REINFORCED PORTLAND CEMENT CONCRETE SURFACES (CONTINUED)

9. LANE-TO-SHOULDER DROPOFF

10. LANE-TO-SHOULDER SEPARATION

| Point No. | Point[1] Distance (meters) | Lane-to-shoulder[2] Dropoff (mm) | Lane-to-shoulder Separation (mm) | Well Sealed (Y/N) |
|---|---|---|---|---|
| 1. | 0. | _ _ _. | _ _ _. | _ |
| 2. | 15.25 | _ _ _. | _ _ _. | _ |
| 3. | 30.5 | _ _ _. | _ _ _. | _ |
| 4. | 45.75 | _ _ _. | _ _ _. | _ |
| 5. | 61. | _ _ _. | _ _ _. | _ |
| 6. | 76.25 | _ _ _. | _ _ _. | _ |
| 7. | 91.5 | _ _ _. | _ _ _. | _ |
| 8. | 106.75 | _ _ _. | _ _ _. | _ |
| 9. | 122. | _ _ _. | _ _ _. | _ |
| 10. | 137.25 | _ _ _. | _ _ _. | _ |
| 11. | 152.5 | _ _ _. | _ _ _. | _ |

Note 1.   Point Distance is from the start of the test section to the measurement location. The values shown are SI equivalents of the 50 ft spacing used in previous surveys.

Note 2.   If heave of the shoulder occurs (upward movement), record as a negative (-) value. Do not record (+) signs, positive values are assumed.

## TABLE OF CONTENTS

## MANUAL FOR DIPSTICK PROFILE MEASUREMENTS

# INTRODUCTION

The Face Construction Technology Dipstick is a manually operated device for the collection of precision profile measurements at a rate and accuracy greater than traditional rod and level survey procedures for individual readings. However, multiple readings may contain systematic cumulative errors, which may cause a shift of the true profile.

The body of the Dipstick houses an inclinometer (pendulum), LCD panels, and a battery for power supply. The sensor of the Dipstick is mounted in such a manner that its axis and the line passing through the contact points of the footpads are co-planar. The sensor becomes unbalanced as the Dipstick is pivoted from one leg to the other as it is moved down the pavement, causing the display to become blank. After the sensor achieves equilibrium, the difference in elevation between the two points is displayed. The Dipstick is equipped with a choice of hardened steel spike feet or ball-and-swivel footpads. The swivel pads should be used on textured pavements.

Calibration of the Dipstick is fixed during manufacture and cannot be altered by the user. Factory calibration accuracy is stated to be 0.038 mm (0.0015 in.) per reading. The user can verify the calibration against a standard calibration block which is provided with the Dipstick.

# OPERATIONAL GUIDELINES

## General Procedures

Dipstick measurements are to be taken by personnel who have been trained in using the device and are familiar with the procedures described in this manual. The detailed scheduling and traffic control at test sites must be coordinated by the regional coordination office contractor (RCOC). All traffic control activities at test sites will be provided by the state or provincial highway agency.

## SHRP Procedures

*Maintenance of Records:* The Dipstick operator is responsible for forwarding all data collected during tests (see forms at the end of this appendix). In addition, the operator is also required to forward other records related to Dipstick operation which are described in the section on record keeping to the RCOC.

*Equipment Repairs:* The RCOC is responsible for ensuring that the SHRP owned equipment is properly maintained. The decisions required for proper maintenance, as well as repair should be made based on the testing schedule and expedited as necessary, to prevent disruption of testing.

*Accidents:* In the event of an accident, the operators will inform the RCOC of the incident as soon as practical. Details of the event shall subsequently be reported in writing to the RCOC to assist in any insurance claim procedures.

## FIELD TESTING

### General Background

The following sequence of field work tasks and requirements provides an overall perspective of the typical work day at a test section.

*Task 1:* Personnel Coordination

   a: Dipstick crew (operator and recorder for manual Dipstick; operator only for auto-read model)

   b: Traffic control crew supplied by the state highway agency (minimum one person or as recommended by the state highway agency)

   c: Other SHRP, State DOT, and RCOC personnel (they are observers and are not required to be present)

*Task 2:* Site Inspection

   a: General pavement condition (within test section limits)

   b: Identify wheel paths

*Task 3:* Dipstick Measurements

   a: Mark wheel paths

   b: Operational checks on Dipstick

   c: Obtain Dipstick measurements

   d: Quality control

*Task 4:* Complete the Dipstick Field Activity Report (see forms at the end of this appendix)

On arrival at the site, the Dipstick operator will carefully plan the activities to be conducted at the site to insure the most efficient utilization of time. While many of the activities can only be accomplished by the Dipstick operator and/or recorder, it will be necessary to enlist the assistance of other personnel at the site to mark the wheel paths. In general, arrangements for this assistance should be made in advance.

Assuming that a manual distress survey is also to be performed at a site, traffic control at a typical GPS site should be available for about eight hours. This should provide adequate time for Dipstick measurements in both the wheel paths as well as the manual distress survey to be completed. As far as the Dipstick productivity rate is concerned, experienced operators have been able to obtain 500 readings per hour.

Collecting profile data is the primary responsibility of the Dipstick operator. In order to ensure that the data collected in all of the SHRP's four regions is identical in format, certain guidelines and standards have been established for the data acquisition and handling phase.

### Site Inspection and Preparation

The pavement must be clear of ice, snow, and puddles of water before profile measurements can be taken with the Dipstick, as such conditions can affect the profile measurements. Pools of water can possibly damage the electronics in the

Dipstick and must be avoided either through adjusting the schedule of profiling trips, or by delaying actual measurements until acceptable conditions exist. If acceptable conditions are present then proceed as follows.

1. Clean both wheel paths of loose stones and debris to prevent slippage of the Dipstick footpads during measurements.

2. Use a chalk line to mark a straight line in each wheel path, 0.8 m (32.5 in.) from the center of the travel lane. The center of the travel lane should be located using the following guidelines.

*Case I:* Where the wheel paths can be easily identified, the center of the travel lane is considered to be midway between the two wheel paths.

*Case II:* Where the wheel paths are not clearly identifiable, but the two lane edges are well defined, the center of the travel lane is considered to be midway between the two lane edges.

*Case III:* Where only one lane edge is well defined, the center of the lane is considered to be 1.8 m (6 ft) from that edge.

If the manual Dipstick is being used, a comment in the data collection sheet should be made on how the center line of the lane was defined. If the auto read Dipstick is used the method of determining the center of the road should be noted in the field notebook. This information will help to collect consistent data in future profile measurements at that section.

3. Lay out and mark straight lines for transverse profile measurements. The lines shall be perpendicular to the edge of the pavement at intervals of 15.25 m (50 ft), starting at Station 0+00 and continuing for 152.5 m (5+00). For GPS sections, eleven lines will be present. The location of lines should be adjusted to avoid raised pavement markings and similar anomalies. The need for and magnitude of such adjustments is to be recorded on the data sheets. Transverse profile measurements are not required for rigid pavements, and in instances where PASCO cross-profile data is available.

### Dipstick Operation for Longitudinal Profile Measurements

#### PRE-OPERATIONAL CHECKS ON THE DIPSTICK
The checks to be performed on the Dipstick prior to testing are described in this section.

1. Check the condition of footpads and replace if necessary with the extra set in the Dipstick case. Clean and lubricate the ball and socket joints on the footpads to insure smooth pivoting of the instrument. When the joint is dirty, pivoting becomes difficult and slippage of the footpad can occur. A cleaning agent such as WD-40 and a light oil for lubrication will work for the ball and socket joint.

2. Install a fresh set of batteries in the instrument and securely close the battery compartment. Batteries should be changed after 4 hours of usage to insure continuity of measurements. Several sets of rechargeable 9 volt batteries should be kept on hand.

3. Check and if necessary, re-tighten the handle on the instrument.

4. Perform the zero check and the calibration check, which are described next. The Zero and Calibration Check Form should be completed whenever these tests are carried out.

*Zero Check:* A zero verification is performed by this test. This test should be performed on a smooth, clean location where the instrument can be properly positioned (the carrying case for the Dipstick, or a flat board will suffice). After positioning the Dipstick draw two circles around the two footpads and note the reading on the display (reading = R1). The instrument should then be rotated 180 degrees and the two footpads placed on the two circles which were drawn earlier. Note the reading obtained (reading = R2). If the readings from the two placements (R1 and R2) add up to within ±0.001 the Dipstick has passed the zero check. If the addition of two readings do not fall within these limits, zero adjustment is necessary. The zero adjustment should be performed using the following procedure.

First obtain the average of the two Dipstick readings:

$$e = 0.5 \ (R1 + R2)$$

Then subtract this value from reading R2 to obtain R2o:

$$R2o = R2 - e$$

Then, with the Dipstick still in the R2 reading position loosen the set screw and adjust the start end adjusting pin up or down so that the display reads R2o. Thereafter, tighten the set screw and rotate the Dipstick back to the R1 reading position and read the display (reading = R1o). Now the addition of R1o and R2o should be within tolerance. If this sum is not within tolerance repeat the adjustment procedure until the two readings are within tolerance.

This zero adjustment is the only adjustment the operator is allowed to make on the Dipstick.

*Calibration Check:* After the zero check and zero adjustments are performed as required, the calibration of the device must be checked. To check the calibration, place the 3.175 mm (0.125 in.) calibration block under one of the Dipstick footpads. The reading displayed, minus 0.125, should equal the previous reading ±0.003. If the answer is not within this tolerance, a SHRP Major Maintenance/ Repair Activity Report should be completed and Face Construction Technology of Norfolk, Virginia should be contacted through the RCOC to repair the Dipstick.

According to the manufacturer the calibration check is needed only if adjustments were required during the zero check. However, for SHRP related measurements both the zero check and calibration check are required at the beginning and end of data collection. Records of these checks should be noted in the Zero and Calibration Check Form shown at the end of this appendix.

### LONGITUDINAL PROFILE MEASUREMENT

To start profile measurements, the Dipstick should be placed on a marked wheel path line at Station 0+00 with the start arrow pointed forward. A clockwise rotation as indicated in Figure 1 should be used in advancing the device. Although the manufacturer does not prohibit a counterclockwise advance, it is prudent to keep the same motion for all test sections so that any potential errors introduced by the rotational direction are consistent in all Dipstick data. As the Dipstick is walked along the marked wheel path, the readings should be recorded on the Longitudinal Profile Data Collection Form if a manual Dipstick is used. If the auto-read Dipstick is used, the readings are stored in the computer attached to the Dipstick. In this case the operator must use the trigger to instruct the computer to store the reading rather than using the automatic storage scheme, to ensure adequate time for the pendulum to stabilize. Use of the automatic mode, even at the lowest production rate, may not allow adequate time for stabilization, thus introducing possible errors in the data.

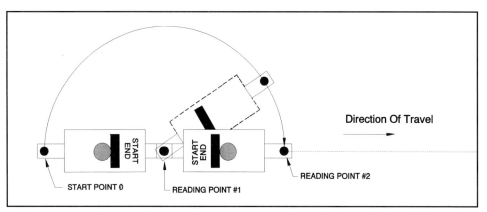

**FIGURE 1**
**Dipstick Operation**

Frequently an operator will introduce bias in the readings by leaning the Dipstick handle laterally from a true vertical position during operation. The operator must not apply any lateral pressure on the handle while the reading is taken. Two runs on each wheel path of the SHRP section must be undertaken by the Dipstick operator, consisting of one run up the wheel path and one run back down the same wheel path. This will accomplish a closed loop survey for each wheel path.

Minor localized cracks and holes in the pavements as well as open joints should be avoided during the Dipstick profile measurement process (e.g., instances where one footpad sinks into a crack or hole or into an open or faulted joint). Also care must be exercised not to place the footpad partially on top of an opening or very wide crack as this may cause slippage during the next advance motion of the Dipstick. If the use of swivel pads is not feasible for some reason, then the operator should avoid isolated depressions by locating the test point on either side of small cracks or holes. If the crack or hole is too big to avoid in this manner, then most likely it should be measured as part of the surface profile and its presence noted in the comments section of the data sheet.

The Dipstick measurements for each wheel path should be taken continuously. However, if for any reason the operator has to stop measurements (e.g., a sudden heavy storm), the point where the last reading was taken must be clearly marked (draw a circle around the footpad). The operator may continue the Dipstick measurements later from this position after placing a comment on the data collection sheet assuming that this point is clearly identifiable on the pavement. The above procedure is suggested by the manufacturer. However, Dipstick measurements on SHRP sections should not be interrupted unless absolutely necessary.

After the forward and the return run have been completed on one wheel path, the post data collection check and quality control check need to be completed before taking measurements on the other wheel path.

### POST DATA COLLECTION CHECK
To ensure the accuracy of the collected data the operator must conduct the zero and calibration checks outlined above after completing measurements on a wheel path. If the Dipstick fails either of these checks, the data must be considered as suspect. If the Dipstick fails the zero check, repeat the profile measurements after adjusting the Dipstick as noted in point 4 above. If the Dipstick fails the calibration test, follow the procedure under point 4 above. If the Dipstick satisfies these checks, proceed with the quality control checks.

## QUALITY CONTROL CHECK

Various forms of error may be introduced into the Dipstick measurements including operator bias, footpad slippage, low battery power, and recording errors. A closed loop survey is used to establish the total accumulated error in the profile measurements acquired with the Dipstick. A closed loop survey is accomplished by completing an initial run along one wheel path and a return run along the exact same wheel path (1 closed loop = 2 runs). The final value of the running sum of the two runs should theoretically result in a zero elevation difference. Any difference in elevation that is obtained is assumed to be due to operator error. By dividing the accumulated error by the length of the loop survey, the error per reading can be obtained. The maximum error allowed is 0.076 mm (0.003 in.) per reading, which corresponds to a total elevation difference of 76 mm (3 in.) for a run 305 m (1000 ft) long (152.5 m up and 152.5 m back; 500 ft up and 500 ft back). If the allowable elevation difference is exceeded during a Dipstick measurement of a test section, the section must be re-tested until acceptable results are obtained.

## Dipstick Operation for Transverse Profile Measurements

### OPERATIONAL CHECKS

The operator should check the equipment using the procedures described above. The checks would include the zero check as well as the calibration check.

### TRANSVERSE PROFILE MEASUREMENT

Dipstick transverse profile measurements will be collected at intervals of 15.25 m (50 ft), starting at Station 0+00. Elevations for each transverse profile location will be measured from the outside edge of the pavement and should extend over the full lane width, with the actual distance depending on lane width and pavement striping. The starting point should be the junction of the transverse measurement line and the inside edge of the white paint stripe along the outside edge of the lane. If no outside edge stripe is present, then the beginning point shall be either the shoulder-lane joint or a point approximately 0.9 m (3 ft) from the center of the outside wheel path. A comment should be entered in the data sheet on how the starting point was determined. The initial elevation is arbitrarily established as "zero" and the subsequent readings are recorded relative to this benchmark. The combination of these measurements provide a measure of the pavement cross slope.

To begin the transverse profile measurements, the Dipstick is placed at the outside edge of the pavement starting at Station 0+00 with the start arrow pointed towards the pavement centerline (see Figure 1). If the manual Dipstick is being used the measurements should be recorded on the Transverse Profile Data Collection Form. If the auto-read Dipstick is used it should be triggered to record the readings. The Dipstick operator should complete two runs per transverse profile of each SHRP section; one run up the transverse line and a return run back along the transverse profile to complete the closed loop survey.

### QUALITY CONTROL FOR TRANSVERSE PROFILE

The total accumulated error in the transverse profile measurement shall be established by a closed loop survey. The forward and return run along a transverse section is utilized to compute this error. The maximum allowable error for the transverse profile measurements is 0.076 mm per reading, or 1.8 mm total (0.003 in. per reading, or 0.072 in. total) for a transverse run 7.32 m (24 ft) long (3.66 m up and 3.66 m back; 12 ft up and 12 ft back). For a transverse run 9.76 m (32 ft) long (4.88 m up and 4.88 m back; 16 ft up and 16 ft back), the maximum allowable total error is 2.4 mm (0.096 in.).

### Data Backup

The importance of safeguarding the Dipstick data cannot be overstated. Backup copies of the Dipstick data must be made without exception after completion of data collection for each test section. Whether the data is recorded on the forms given at the end of this appendix, or on electronic media, copies should be made at the earliest time possible.

A minimum of two complete copies shall be made of all Dipstick data. One copy will be transmitted by mail to the regional coordination office while the second will be retained by the operator in case the first copy is lost in the mail.

## CALIBRATION

### General Background

Internal calibration of the Dipstick is fixed during manufacture and cannot be altered by the user. However, to ensure that the instrument is operating properly the calibration can be checked using the zero check and the calibration check described under Pre-operational Checks on the Dipstick. If the Dipstick fails the calibration test it should be returned to the manufacturer for repair.

The gage block used during the calibration check should be calibrated to an accuracy of $3.175 \pm 0.025$ mm ($0.125 \pm 0.001$ in.) using a local calibration laboratory or a calibration micrometer to ensure the minimum 4:1 ratio of accuracy of gage blocks to the Dipstick. The gage block should be recalibrated once every year, or more frequently, depending on (1) presence of oxidation, (2) evidence of erosion, and (3) possible damage caused by accidental mishandling in the field.

If the calibration block (gage block) thickness is not within $3.175 \pm 0.025$ mm ($0.125 \pm 0.001$ in.), all data collected since the last check of the block are suspect and may have to be disregarded.

### Calibration Frequency

The zero check and the calibration check should be conducted by the operator prior to and after any Dipstick measurements. If the Dipstick fails the calibration test, the approval from an RCOC engineer is required before shipping the equipment to the manufacturer.

## EQUIPMENT MAINTENANCE AND REPAIR

### General Background

Scheduled preventive maintenance will serve as a means of ensuring proper operation of the equipment as well as identifying potential problems. Timely identification of problems will help to avoid costly delays or incomplete data which could result from on site equipment malfunction. The time constraints on the profile testing program require that maintenance activities be performed prior to mobilization for testing. During testing it is necessary that the operator be constantly aware of the proper functioning of the equipment. There will be little time to accomplish more than the required initial checks at the site in preparation for the test day. Therefore, preventive maintenance must be performed as a routine function at the end of each test/travel day and on days when no other work is scheduled.

Minimizing the rate of equipment deterioration is the responsibility of the RCOC and individual operators. Specific, detailed maintenance procedures are contained in the manuals provided with each individual piece of equipment, and the operator must become intimately familiar with the maintenance recommendations contained in all equipment manuals. This section is intended to reinforce the concept of maximum equipment dependability, which is critical to the effectiveness of the LTPP program.

It is not the intention of this guide to supersede manufacturers' minimum services and service intervals, but to provide supplementary service requirements. Where there is a conflict between this guide and the manufacturers' instructions, the more stringent requirements should be followed.

## Routine Maintenance

Routine maintenance includes those functions which can be easily performed by the operator with minimal disassembly of a particular device. Routine maintenance for the Dipstick includes the cleaning and lubrication of the ball and socket joints on the footpads, replacement of the batteries and cleaning of the battery contacts. These are basic and easily performed preventive measures and should always be completed prior to operation of the equipment.

The following list of pre-operation preventive maintenance items is not complete, but is intended to show the extent and detail required before the operation checks are performed. This list of items is not to supersede manufacturers' minimum requirements for warranty compliance.

1. *Exterior:* Check general appearance, glass display (should be clean), ball and socket joint for the footpads (should be properly lubricated).

2. *Accessories:* Be sure adequate supplies of consumables are on hand (e.g., batteries, grease, WD-40).

In the Dipstick Field Activity Report the first line of information following the section identification data is an acknowledgement that the pre-operation checks were properly performed.

## Scheduled Major Maintenance

Scheduled major preventive services shall include much more than the routine checks and will require some disassembly of the equipment by personnel with technical capabilities beyond the skill of the operators or RCOC staff. The SHRP Major Maintenance/Repair Report should be used by the operator to report the performance of necessary services. This form will also serve to inform the RCOC of the condition of the Dipstick on a regular basis. Items such as battery connecter replacement would fall into the major maintenance category. The appropriate service intervals are outlined in the equipment manufacturer's manual.

## Equipment Problems and Repairs

Regardless of the quality of the preventive program there will probably be equipment failures during the LTPP program. When these occur it is extremely important that repairs or replacement of items be accomplished in a timely fashion. During periods when there is no scheduled testing, these problems are easily handled. However, if they occur during mobilization or while on-site, significant problems in scheduling and coordination could develop. To help minimize the impact of equipment problems it is essential that the operator immediately notify the RCOC and any other agencies or individuals as necessary. The responsibility for equipment maintenance and repair activity rests with each RCOC. However, the RCOC should keep LTPP staff informed of any major problems concerning the equipment. When repairs are necessary and must be performed by an outside agency, the operator will report this information on the SHRP Major Maintenance Report form as an unscheduled maintenance activity. Details of the circumstances during field testing related to this maintenance activity should be noted on the daily activity report. Minor repairs performed by the operator at times other than during testing should be recorded on a daily activity report that clearly states no testing was performed. In this case, no reference information (section number, etc.) will be entered.

## RECORD KEEPING

The Dipstick operator will be responsible for maintaining the following forms and records:

1. Dipstick Field Activity Report

2. Major Maintenance/Repair Report

3. Zero and Calibration Check Form

All of these forms are included in at the end of this appendix.

Each of these records shall be kept in up-to-date files by each RCOC with one complete set kept on file at the regional office. A description of each of these forms follows.

### Dipstick Field Activity Report

The Dipstick Field Activity Report is an document prepared by the Dipstick operator which includes a commentary of all activities associated with profile measurements. The information to be noted in this form includes: the section information, time to complete all measurements, any downtime, information related to productivity, any factors which might affect the collected test data and names and organizations of other personnel present at the site. The names of these personnel would be invaluable if an accident occurs at the test site. A space is provided in this form for the operator's acknowledgment that pre-operation checks were conducted prior to any Dipstick testing. The operator should fill out a Dipstick Field Activity Report at every section where a Dipstick test is conducted. In addition this form should be completed whenever the operator performs maintenance on the Dipstick. The original of this report should be kept by the operator with a copy forwarded to the RCOC.

### SHRP Major Maintenance/Repair Report

When any major maintenance or repair must be performed by an outside agency, the SHRP Major Maintenance/Repair Activity Report must be filed. Routine maintenance (pre-operation checks) and minor, operator performed repairs should be reported on the Field Activity Report.

### Zero and Calibration Check Form

This form should be completed whenever the zero and calibration checks are carried out.

These reports and forms may be photocopied from this book for use in the field.

The Dipstick Field Activity Report, the FHWA-LTPP Major Maintenance/Repair Activity Report, and the Zero Check/Calibration Check forms are used to monitor the performance of the device itself.

The data collection forms for transverse and linear profiles are used to record data from the pavement section.

**Blank Reports and Forms for Dipstick Profile Measurements**

**DIPSTICK FIELD ACTIVITY REPORT**

SHRP REGION_____STATE CODE_____SHRP ASSIGNED ID_____
STATE_____TESTING_____DISTRICT_____
LTPP EXPERIMENT CODE_____ROUTE/HIGHWAY NUMBER_____
   EQUIPMENT SERIAL NUMBER_____
   TESTING DATE_____ SHEET NUMBER_____
   WEATHER_____

=========================================================================

   DIPSTICK PRE-OPERATION CHECKS_____ (initials)

                                           TIME

                READY TO TEST     _____
                BEGIN TESTING    _____
                END TESTING      _____
                START TRAVEL     _____
                END TRAVEL       _____

DOWN TIME _____ HOURS

REASONS_____
_____
_____

ADDITIONAL REMARKS

_____
_____
_____
_____

DIPSTICK PROFILE CREW                       TRAFFIC CONTROL CREW
NAMES:_____                  NAMES:_____
         _____                        _____
         _____                        _____
                                         _____
                                         _____

COPIES:  RCOC                                FORM F01/SEPT 1990

# SHRP MAJOR MAINTENANCE/REPAIR ACTIVITY REPORT

REGION_____DATE_____

EQUIPMENT ID
     MAKE:<u>FACE CONSTRUCTION TECHNOLOGY, INC.</u>   MODEL:_____

SERIAL NUMBER:_____

REASON FOR MAINTENANCE WORK (CHOOSE ONE ONLY)

     SCHEDULED_____       NONSCHEDULED_____

DESCRIPTION OF MAINTENANCE AND REASON:

_____
_____
_____
_____
_____
_____
_____

AGENCY PERFORMING MAINTENANCE                  COST:_____

     NAME:_____
     STREET ADDRESS:_____
     CITY:_____
     PHONE NUMBER:_____
     CONTACT NAME:_____
     DATE IN:_____
     DATE OUT:_____

===========================================================================

COPIES TO:  RCOC, LTPP DIVISION

FORM F02/DEC 1992

ZERO CHECK

First Reading_____

Rotate 180 degrees

Second Reading_____

    Total, if within ±0.001 proceed or else adjust the start end pin as suggested in the manual and repeat the zero check.


CALIBRATION CHECK

First Reading_____

Place calibration block

Second Reading_____  -  0.125 = First Reading ±0.003 proceed or else contact FACE through RCO

## MANUAL DIPSTICK DATA COLLECTION FORM
### (Longitudinal Profile)

TEST SITE:_____ DATE:_____ WEATHER:_____

OPERATOR:_____ RECORDER:_____ DIPSTICK SERIAL NUMBER:_____

START TIME:_____ STOP TIME:_____ WHEELPATH:_____

| Distance (ft) | Elevations (in.) | | Distance (ft) | Elevations (in.) | | Distance (ft) | Elevations (in.) | | Distance (ft) | Elevations (in.) | |
|---|---|---|---|---|---|---|---|---|---|---|---|
| | Pass 1 | Pass 2 | | Pass 1 | Pass 2 | | Pass 1 | Pass 2 | | Pass 1 | Pass 2 |
| 1 | | | 26 | | | 51 | | | 76 | | |
| 2 | | | 27 | | | 52 | | | 77 | | |
| 3 | | | 28 | | | 53 | | | 78 | | |
| 4 | | | 29 | | | 54 | | | 79 | | |
| 5 | | | 30 | | | 55 | | | 80 | | |
| 6 | | | 31 | | | 56 | | | 81 | | |
| 7 | | | 32 | | | 57 | | | 82 | | |
| 8 | | | 33 | | | 58 | | | 83 | | |
| 9 | | | 34 | | | 59 | | | 84 | | |
| 10 | | | 35 | | | 60 | | | 85 | | |
| 11 | | | 36 | | | 61 | | | 86 | | |
| 12 | | | 37 | | | 62 | | | 87 | | |
| 13 | | | 38 | | | 63 | | | 88 | | |
| 14 | | | 39 | | | 64 | | | 89 | | |
| 15 | | | 40 | | | 65 | | | 90 | | |
| 16 | | | 41 | | | 66 | | | 91 | | |
| 17 | | | 42 | | | 67 | | | 92 | | |
| 18 | | | 43 | | | 68 | | | 93 | | |
| 19 | | | 44 | | | 69 | | | 94 | | |
| 20 | | | 45 | | | 70 | | | 95 | | |
| 21 | | | 46 | | | 71 | | | 96 | | |
| 22 | | | 47 | | | 72 | | | 97 | | |
| 23 | | | 48 | | | 73 | | | 98 | | |
| 24 | | | 49 | | | 74 | | | 99 | | |
| 25 | | | 50 | | | 75 | | | 100 | | |
| TOTAL | | | | | | | | | | | |

COMMENTS: _____
_____
_____

DATE: _____

OPERATOR: _____

RECORDER: _____

DIPSTICK SERIAL NUMBER: _____

STATE CODE: __ __

SHRP SECTION I.D.: __ __ — __ __

| LOCATION | DIPSTICK READING | | | | | | | | | | | | | | | TOTAL | SUM |
|---|---|---|---|---|---|---|---|---|---|---|---|---|---|---|---|---|---|
| | 1 | 2 | 3 | 4 | 5 | 6 | 7 | 8 | 9 | 10 | 11 | 12 | 13 | 14 | 15 | | |
| 0+00 | | | | | | | | | | | | | | | | | |
| 0+50 (15 m) | | | | | | | | | | | | | | | | | |
| 1+00 (30 m) | | | | | | | | | | | | | | | | | |
| 1+50 (45 m) | | | | | | | | | | | | | | | | | |
| 2+00 (60 m) | | | | | | | | | | | | | | | | | |
| 2+50 (75 m) | | | | | | | | | | | | | | | | | |
| 3+00 (90 m) | | | | | | | | | | | | | | | | | |
| 3+50 (105 m) | | | | | | | | | | | | | | | | | |
| 4+00 (120 m) | | | | | | | | | | | | | | | | | |
| 4+50 (135 m) | | | | | | | | | | | | | | | | | |
| 5+00 (150 m) | | | | | | | | | | | | | | | | | |

FORM S1/December 1992

## TABLE OF CONTENTS

**C**

**MANUAL
FOR
FAULTMETER
MEASUREMENTS**

# INTRODUCTION

### Measurement of Faulting in the LTPP Program

This manual is intended for use by the FHWA-LTPP Regional Coordination Office personnel and others responsible for using the Faultmeter to measure faulting of jointed concrete pavements, and for measuring lane-to-shoulder dropoff on LTPP pavement test sites.

The change in joint faulting and lane-to-shoulder dropoff with time are important indicators of pavement performance. The Digital Faultmeters will be used to collect this data. It is the responsibility of each regional coordination office contractor to store, maintain, and operate their Faultmeter for faulting and lane-to-shoulder dropoff data collection.

### The Georgia Digital Faultmeter

The electronic Digital Faultmeter was designed to simplify measuring concrete joint faulting. This meter was designed, developed and built by the Georgia Department of Transportation (Georgia DOT) Office of Materials and Research personnel.(1) The Faultmeter is very light and easy to use. The unit, shown in Figure 1, weighs approximately 3.2 kg (7 lbs) and supplies a digital readout with the push of a button located on the carrying handle. It reads out directly in millimeters (e.g., a digital readout of "6" indicates 6 mm (0.25 in.) of faulting) and shows whether the reading is positive or negative. The unit reads out in 1 second and freezes the reading in display so it can be removed from the road before reading for safer operation. The legs of the base of the Faultmeter are set on the slab in the direction of traffic on the "leave side" of the joint. The measuring probe contacts the slab on the approach. Movement of this probe is transmitted to a Linear Variance Displacement Transducer (LVDT) to measure joint faulting. The joint must be centered between the guidelines shown on the side of the meter.

Any slab which is lower on the leave side of the joint will register as a positive faulting number. If the slab leaving the joint is higher, the meter gives a negative reading.

The amount of time it takes to complete the faulting survey of a LTPP test section depends on the number of joints and cracks encountered and on the amount of time needed to measure and record the location of each joint and crack. Generally, it should take less than 30 minutes to measure and record faulting and lane-to-shoulder dropoff on a 150 m (500 ft) test section using this device.

# OPERATING THE FAULTMETER

This section gives step-by-step operating instructions. The Faultmeters manufactured for SHRP by the Georgia DOT have several unique features, which have been added to simplify their operation, increase range of measurement to 22 mm (0.95 in.), and increase their "reach" to 100 mm (4 in.) to allow for spanning spalls and excess joint material on the slab surface. The handle is removable for facilitating packaging when traveling from one site to another. The handle (made from PVC) is equipped with a threaded base, nut, and washer for attachment to the base. To attach the handle, remove the washer, insert the threaded end of the handle through the hole in the base and screw the nut and washer on the bottom side of the base. To initiate testing, plug in the switch wire from the handle to the connector on the Faultmeter. The readout display will remain blank until the test button on the grip of the handle has been depressed.

Grip the handle of the meter with the thumb resting lightly on the test button. Use the right hand when testing the outside lane. This allows the operator to stand safely on the shoulder, facing traffic, while making the test. There is an arrow on the meter showing traffic direction. Set the meter on the leave side of the joint. A probe contacts the slab on the approach side. The joint must be approximately centered between the two marks on each side of the meter. Depress and instantly release the test button. A 1-second tone will sound (the word "hold" is also displayed in the lower left hand corner adjacent to the reading taken). As soon as the tone stops, lift the meter and move away from the pavement. The reading will remain "frozen" until the test button is depressed again. This feature allows the operator to move away from traffic before the meter is read.

As indicated in Chapter 3 of the Data Collection Guide (2), faulting of transverse joints and cracks is measured as the difference in elevation to the nearest 1 mm (0.04 in.) between the pavement surface on either side of a transverse joint or crack. It is measured 0.3 m (1 ft) and 0.76 m (2.5 ft) from the outside slab edge. Where the Faultmeter does not span existing spalling or other anomalies, the meter should be offset to avoid including such anomalies in the readings. Measurements are taken at every joint and crack. This data is to be recorded on Distress Survey Sheet 6. If more than 27 joints or cracks have faulting, record the measurements on additional copies of Sheet 6. The distance from the start of the test section to the point where the measurement is taken is also recorded. This distance may be obtained with measuring wheel, steel tape, or counting slabs, if joint spacing is known. Faulting is assumed to be positive. Therefore,

**FIGURE 1**
**The Georgia Digital Faultmeter in Use**

the space to the left of the entry of measured faulting is to be filled with a negative sign when necessary. If the approach slab is higher than the departure slab, no positive sign is to be entered. If the approach slab is lower, a negative sign is entered. The readings recorded on the Faultmeter are reported in millimeters on Sheet 6.

Lane-to-shoulder dropoff is measured as the difference in elevation to the nearest 1 mm (0.04 in.) between the pavement surface and the adjacent shoulder surface. Measurements are taken at the beginning of the test section and at 15 m (50 ft) intervals (a total of six measurements) at the lane/shoulder interface or joint. Lane-to-shoulder dropoff typically occurs when the outside shoulder settles. However, heave of the shoulder may occur due to frost action or swelling soil. If heave of the shoulder is present, it should be recorded as a negative value. At each point where there is no lane-to-shoulder dropoff, enter "0." This data is again to be entered on Distress Survey Data Sheet 3 for Asphalt Concrete-Surfaced Pavements, Data Sheet 7 for Jointed Concrete Pavements, and Data Sheet 10 for Continuously Reinforced Concrete Pavements.

The distance from the center of the measuring probe to the edge of the forward foot of the base is 100 mm (4 in.) to allow easy placement on the joint and for more overhang, to measure shoulder dropoff. In addition, the base feet are 50 mm (2 in.) long in an attempt to bridge any bad crack or low spot in the pavement. The Faultmeters will read up to 22 mm (0.95 in.). Differential elevations greater than 22 mm will still need to be measured using the machined spacer block supplied with the Faultmeter.

The meter is equipped with an automatic shutoff. The display will go blank (turning the system off) approximately 15 minutes after the last test has been run (15 minutes from the last time the test button has been depressed).

## CALIBRATION

Although the meter is very stable, it should be checked at the beginning and end of every day to assure correct readings. Calibration is performed by setting the meter on the calibration stand, which has been provided with the Faultmeter. The front end of the Faultmeter is to be aligned with the calibration "20" mark. In this position, the probe rests on a 20 mm (0.75 in.) block. A reading of 20 should be obtained. The meter should then be lined up with the 0 mark and display a reading of 0.

As long as the 0 and 20 readings are obtained, the unit is working properly. If not, discontinue testing and reset the calibration as described below. Be sure to check for any electronic malfunction before adjusting the calibration and 0 controls. Extremely weak batteries could also cause an erroneous reading.

The following set of procedures should be needed only after correcting an electronic malfunction (initial calibration has already been performed on each of the units). Zero and calibration controls are provided on the control PC board. Both are 20-turn potentiometers. Before beginning, turn each approximately 10 turns from either end to center the adjustment. Set the Faultmeter on the calibration stand in the 0 position. Hold the test button down for the following steps: 1) Loosen the bolt that clamps the LVDT in its holder, 2) Slide the LVDT up or down to obtain an approximate 0 reading, and 3) Retighten the LVDT clamp. All other adjustments will be made with the zero and calibration potentiometers.

Adjust the 0 control to exactly 0. The easiest way is to slowly turn the control until the minus (-) sign in the display flickers on and off. Next, move the Faultmeter up on the calibration block to the "20" mark. A reading of 20 should be obtained.

The easiest way is to slowly turn the control up to the next higher number (21 in this case) and down to the next lower number (19 in this case). Turn the control exactly halfway between these limits to be in the center of the desired number (20 in this instance).

Recheck the Faultmeter on the calibration stand in both the zero (0) and calibrate positions. Touch up either control slightly, if necessary. It is a good idea to put a drop of fingernail polish or other adhesive on the adjusting screw of the controls after setting them. This keeps the controls from accidentally moving because of vibration. Try to leave the screwdriver slot open in case they should ever have to be adjusted again.

## MAINTENANCE

The only maintenance normally required for the Faultmeter would be for replacing batteries. When the batteries need replacement, a 1-second continuous tone will be followed by a 3-second pulsating tone and "lo batt" will show in the upper left corner of the display. The readings will still be accurate, but the batteries should be replaced within a day or so. The unit takes eight AA batteries. As previously noted, an automatic cutoff switch has been provided so that the meter will turn on the first time the test button is depressed and the meter will turn off approximately 15 minutes from the last time the test button is depressed. This will help to minimize wear on the batteries.

These meters have also been equipped with a reverse polarity protection fuse. If the batteries are installed backwards, the fuse will blow out, but the meter electronics will not be damaged. It is a good idea to remove the fuse while the batteries are being replaced, so it will not be blown if the battery plug is touched to the battery pack backwards.

If the measuring rod does not move freely, the readings will be in error. This should not be a problem, as the rod is made of stainless steel and will not rust. If the rod should get coated with road film and dust, it may be cleaned with a damp cloth. Do not clean with penetrating oil or any products that will leave an oily residue, as this will cause dust to adhere to the rod. If the rod "clicks" when the meter is lifted from the pavement, this is a good indication that it is sliding freely.

## REFERENCES

1. Jerry Stone, Georgia Digital Faultmeter, Report FHWA-GA-91-SP9010, Federal Highway Administration, January 1991.

2. Data Collection Guide for Long Term Pavement Performance Studies, Operation Guide SHRP-LTPP-OG-001, Strategic Highway Research Program, January 1990.

Appendix C